Neurocriminology

Neurocriminology
Forensic and Legal Applications, Public Policy Implications

Diana M. Concannon

CRC Press
Taylor & Francis Group
Boca Raton London New York

CRC Press is an imprint of the
Taylor & Francis Group, an **informa** business

CRC Press
Taylor & Francis Group
6000 Broken Sound Parkway NW, Suite 300
Boca Raton, FL 33487-2742

© 2019 by Taylor & Francis Group, LLC
CRC Press is an imprint of Taylor & Francis Group, an Informa business

No claim to original U.S. Government works

Printed on acid-free paper

International Standard Book Number-13: 978-1-138-63280-6 (Hardback)

This book contains information obtained from authentic and highly regarded sources. Reasonable efforts have been made to publish reliable data and information, but the author and publisher cannot assume responsibility for the validity of all materials or the consequences of their use. The authors and publishers have attempted to trace the copyright holders of all material reproduced in this publication and apologize to copyright holders if permission to publish in this form has not been obtained. If any copyright material has not been acknowledged please write and let us know so we may rectify in any future reprint.

Except as permitted under U.S. Copyright Law, no part of this book may be reprinted, reproduced, transmitted, or utilized in any form by any electronic, mechanical, or other means, now known or hereafter invented, including photocopying, microfilming, and recording, or in any information storage or retrieval system, without written permission from the publishers.

For permission to photocopy or use material electronically from this work, please access www.copyright.com (http://www.copyright.com/) or contact the Copyright Clearance Center, Inc. (CCC), 222 Rosewood Drive, Danvers, MA 01923, 978-750-8400. CCC is a not-for-profit organization that provides licenses and registration for a variety of users. For organizations that have been granted a photocopy license by the CCC, a separate system of payment has been arranged.

Trademark Notice: Product or corporate names may be trademarks or registered trademarks, and are used only for identification and explanation without intent to infringe.

Library of Congress Cataloging-in-Publication Data

Names: Concannon, Diana M., author.
Title: Neurocriminology : forensic and legal applications, public policy
implications / Diana Concannon.
Description: Boca Raton, FL : CRC Press, [2019] | Includes bibliographical
references and index.
Identifiers: LCCN 2018018719| ISBN 9781138632806 (hardback : alk. paper) |
ISBN 9781315208039 (ebook)
Subjects: LCSH: Criminology—Government policy. | Criminology. | Forensic
sciences.
Classification: LCC HV6025 .C588 2019 | DDC 364.3—dc23
LC record available at https://lccn.loc.gov/2018018719

Visit the Taylor & Francis Web site at
http://www.taylorandfrancis.com

and the CRC Press Web site at
http://www.crcpress.com

Printed and bound by CPI Group (UK) Ltd, Croydon, CR0 4YY

Can brain scans be used to determine whether a person's inclined toward criminality or violent behavior?

You will rule on that.

Then Senator Joseph Biden
Supreme Court nomination hearing of John Roberts

Contents

Author		**xiii**
Introduction		**xv**

1   Neurocriminology: How Did We Get Here? A Brief History of Criminology   1

Introduction	1
The Pre-Classical Period: The Devil Made Him (or Her) Do It	2
The 18th Century and the Classical School	2
The 19th Century: Phrenology, the Positivists, and Lombroso	5
The 20th Century: Mainstream Criminology, Critical Criminology, Left Realism, and Conservative Criminology	9
Mainstream Criminology	10
Control Theories	13
The 21st Century: Precursors to Neurocriminology	20
Key Terms	23
Use Your Brain	24
Test Your Knowledge	24
Apply Your Knowledge	25
Bibliography	26

2   Brain Basics: How Neurocriminology Is Possible   29

Introduction	29
The "Neuro" of Neurocriminology	34
The Structure and Functions of the Brain	38
The Cerebral Cortex	38
Occipital Lobes	39
Temporal Lobes	40
Parietal Lobes	40
Frontal Lobes	41
The Limbic System	42
Hypothalamus	42
Hippocampus	42

| | viii | Contents |

Amygdala	43
Cingulate Cortex	43
Brodmann's Areas	44
Common Causes of Brain Dysfunction	45
Tumor	45
Stroke	46
Trauma	47
Epilepsy	49
Neurocognitive Disorders	50
Psychotic and Mood Disorders	51
Personality Disorders	52
Substance Abuse	53
Key Terms	55
Use Your Brain	57
Test Your Knowledge	57
Think About It	59
Bibliography	59

3 Overview of Advances in Neuroimaging 63

Introduction	63
How Neuroimaging Works	65
Structural Neuroimaging Techniques	66
Computed Tomography/Computerized Axial Tomography	66
Magnetic Resonance Imaging	66
Functional Neuroimaging Techniques	67
Electroencephalography/Evoked Potentials and Magnetoencephalography/Magnetic Source Imaging	68
Functional Magnetic Resonance Imaging	70
Positron Emission Tomography	71
Single-Photon Emission Computerized Tomography	71
Strengths and Limitations of Neuroimaging	72
Reliability	73
Validity	73
Inferential Gaps	73
Causation	74
Group-to-Individual Problem	74
Key Terms	75
Use Your Brain	75
Test Your Knowledge	75
Apply Your Knowledge	76
Bibliography	77

Contents ix

4 Neurocriminology: Preliminary Applications 79

Introduction	79
Charles Joseph Whitman	80
Jack Ruby	82
John Hinckley, Jr.	83
Vincent Gigante	87
Key Terms	90
Use Your Brain	90
Test Your Knowledge	90
Apply Your Knowledge	91
Bibliography	92

5 Criminals in the Lab 93

Introduction	93
Murderers	94
Predatory versus Affective Murderers	95
Murderers with Severe Mental Illness	97
Adolescent Murderers	98
Non-Homicidal Offenders	98
Sexual Offenders	101
Rapists	101
Pedophiles	104
White-Collar Criminals	105
Criminal Subtype Conclusions	105
Key Terms	106
Use Your Brain	106
Test Your Knowledge	106
Apply Your Knowledge	108
Bibliography	108

6 Neurocriminology in the Criminal Justice System: Prevention and Investigation 111

Prevention: The Brain Can Change	112
Roper v. Simmons	114
Graham v. Florida	117
Miller v. Alabama	119
Prevention and Intervention	121
The Brain and Early-Onset Antisocial Behaviors	122
Prevention and Early Intervention Efforts	124
Drug Courts	125

x   Contents

The Brain and Drug Abuse   125
Preliminary Findings   126
Veterans Treatment Courts   130
Neurocriminology and Specialty Courts: Possibilities and
Limitations   136
Neuroscience and Criminal Investigation   137
Witness Memory Recovery   144
Lie Detection   146
Key Terms   152
Use Your Brain   153
Test Your Knowledge   153
Apply Your Knowledge   154
Bibliography   155

7   Neurocriminology in the Criminal Justice System: Prosecution and Sentencing   159

Introduction   159
Competency to Stand Trial   160
Guilt   164
Sentence Mitigation   172
The Sentencing of Brian Dugan   176
State of Florida v. Grady L. Nelson   179
Ineffective Use of Counsel   180
Glenn Douglas Anderson v. Marty Sirmons   180
Key Terms   181
Use Your Brain   182
Test Your Knowledge   182
Apply Your Knowledge   183
Bibliography   183

8   Neuroscience and Law: International Applications   185

Introduction   185
England and Wales   186
Canada   187
The Netherlands   187
India   189
Italy   190
Use Your Brain   192
Test Your Knowledge   192
Apply Your Knowledge   193
Bibliography   194

9 Neurocriminology: Present Context and Possible Future 195

Bibliography 197

Index 199

Author

Diana M. Concannon, PsyD, PCI, is currently Associate Provost for Strategic Initiatives and Partnerships; Dean, California School of Forensic Studies (CSFS); and Director, California Psychology Internship Consortium (CPIC) at Alliant International University, based in San Diego, California. Dr. Concannon has worked at Alliant since 2009 and overseen the establishment and set the curriculum for six California-based PsyD programs in Clinical Forensic Psychology, a PhD program in Clinical Forensic Psychology, a PhD program in Psychology, Public Policy & Law, a blended master's program in Applied Criminology, and an online master's program in Applied Criminology. She has taught courses on Advanced Theories of Personality, Forensic Ethics, Foundational Concepts of Victimology I: Victimology Investigations, Victimology Intervention, and Understanding the DSM-5. She is presently developing new programs to advance the university's mission and goals, including a Master's of Science Forensic Behavioral Science, a Master's of Science Forensic Leadership and Administration, and several Advanced Practice Certificates to support emergency responders. She is a forensic psychologist, licensed to practice in California, New York, Utah, and Washington, DC; a Professional Certified Investigator by the American Society for Industrial Security; a Rape Escape Instructor; and a Loyola Law School trained Mediator. She is the author of *Kidnapping: An Investigator's Guide to Profiling* (2008) and *Kidnapping: An Investigator's Guide,* Second Edition (2013), both with Elsevier.

Introduction

In December 2017, the brain of Stephen Paddock, the 64-year-old gunman responsible for the deadliest mass shooting in U.S. history, was sent to Stanford University for study.

The ongoing investigation had yet to reveal why Paddock checked into a Las Vegas hotel on October 1, 2017 and opened fire from his 32nd story room, killing 58 and wounding more than 500 of the 22,000 who attended the concert below. Connections to terrorism or another shooter were ruled out early. Involvement of Paddock's girlfriend, who Paddock had sent to her native country prior to the attack, was also ruled out. Paddock reportedly had no criminal history, was a retired accountant, and was financially secure. He was a high-stakes poker player. Some interviewed after the event described him as having "a God complex." He had rented hotel rooms above music venues in other localities, but the significance of this (if any) was unclear. He was able to purchase firearms without raising any red flags; in addition to the 23 weapons in his hotel suite, Paddock had more than 50 pounds of exploding targets and 1,600 rounds of ammunition in the car he parked in the lot of the hotel.

Prior to accessing the additional weaponry, Paddock died from a self-inflicted gunshot wound to the mouth. A visual inspection of Paddock's brain found "no abnormalities," but cause and manner of death were pended until the findings of Stanford's multiple forensic analyses, including a neuropathological examination of Paddock's brain, were completed.

Any findings from these more in-depth analyses would clearly not alter the tragedy or its heartbreaking aftermath.

Yet absent revelations regarding other criminogenic factors—a past history of violence, a history of trauma or abuse, substance abuse—a finding of a brain abnormality might provide some insight into Stephen Paddock's otherwise inexplicable act. Conversely, an absence of brain abnormality would serve as a reminder that the current state of science cannot always illuminate the reasons why a specific individual becomes a violent criminal.

As a forensic psychologist, I have supported the judicial system to render decisions at various stages in the criminal justice process, from determining competency to stand trial, to assessing risk of recidivism, to deciding if or under what conditions a convicted criminal should be released back into society. These recommendations are informed by clinical judgment and

xvi   Introduction

the use of standardized assessment instruments that consider actuarial data found to correlate with the specific forensic question at hand. The assessments invariably include inquiries into, among other factors, substance abuse, brain injuries, and cognitive functioning, each of which can provide insight into an individual's past criminal behavior and likely future actions.

By translating neuroscience—the study of the brain and nervous system—to antisocial behavior, *neurocriminology* adds to our knowledge of the factors associated with criminality. It is a word first coined by Canadian criminologist James Hilborn, and defined by University of Pennsylvania Professor of Criminology, Psychiatry, and Psychology Adrian Raine as "the application of the principles and techniques of neuroscience to understand the origins of antisocial behavior" to "improve our ability to prevent the misery and harm crime causes."

Given this text's primary focus on the application of neuroscience to the criminal justice system—as opposed to the civil court system, where it is also making contributions—the term *neurocriminology* has been adopted.

The application of neuroscience to crime and crime control has generated both interest and controversy in a relatively short span of time.

During the 1990s, the potential impact of neuroscience in the courtroom entered the broader public awareness during the murder trial of Herbert Weinstein, the 65-year-old New York advertising executive with no prior criminal history who strangled his wife and threw her body from their 12th floor apartment in an attempt to stage his murderous act as a suicide. Weinstein's attorneys petitioned the court to introduce positron emission tomography (PET) scans that revealed that Weinstein had an arachnoid cyst on his brain, resulting in the displacement of the left frontal lobe. This, the defense contested, caused Weinstein to lack criminal responsibility for killing his wife due to mental disease or defect as, at the time of the offense, "he lacked the cognitive ability to understand the nature and consequences of his conduct or that his conduct was wrong." This assertion was made based, in part, on the functioning associated with the frontal lobe, which includes emotional regulation and decision-making.

The judge ruled that the PET images were admissible, but that testimony that arachnoid cysts cause violence lacked scientific validity and was, therefore, inadmissible. Weinstein ultimately accepted a plea of manslaughter.

Although Weinstein's PET scan evidence was never introduced at trial, the motion to do so catalyzed significant and important dialogue about the role of neuroscience and brain functioning outside of the courtroom.

In 1996, for example, University of Pennsylvania professor and legal scholar Stephen J. Morse pointed out that the lack of scientifically valid causation between Weinstein's brain condition and violence was the relevant legal issue: "Even a highly abnormal cause will not excuse unless it produces an excusing condition." Morse continued,

Introduction

xvii

> ... emotional states [resulting from the presence of a brain tumor] surely make it harder for any agent to fly straight in the face of other criminogenic variables, such as provocation or stress, but per se they don't render an agent irrational. Other agents may be equally irritable or labile as the result of environmental variables, such as the loss of sleep and stress associated with, say, taking law exams or trying an important, difficult, lengthy case. But these people would not be excused if they offended while in an uncharacteristic emotional state unless that state sufficiently deprived them of rationality.

His perspective was shared by others in the field, and indeed by the courts themselves; neuroscientific evidence played an insignificant role in the criminal justice system for the remainder of the decade.

A modest paradigm shift began in the 2000s.

Vermont Law School professor and legal scholar Oliver Goodenough published an article proposing a series of neural experiments to "provide the opportunity to revisit classic questions at the foundation of legal reasoning," and, specifically, to determine if human moral and legal reasoning represent different mental processes and could or should inform each other.

Additional articles published during the decade, such as those by neuroscientist Robert Sapolsky, and by psychologists Joshua Greene and Jonathan Cohen, asserted that neuroscience has potentially significant relevance to decision-making within the criminal courts. Greene and Cohen went so far as to suggest that:

> New neuroscience will change the law, not by undermining its current assumptions, but by transforming people's moral intuitions regarding free will and responsibility. This change in moral outlook will result not from the discovery of crucial new facts or clever new arguments, but from a new appreciation of old arguments, bolstered by vivid new illustrations provided by cognitive neuroscience. We foresee, and recommend, a shift away from punishment aimed at retribution in favor of a more progressive, consequentialist approach to the criminal law.

In 2007, the John D. and Catherine T. MacArthur Foundation announced an initial $10 million grant to fund the Law and Neuroscience Project, an interdisciplinary effort dedicated to assisting "the criminal justice system understand neuroscientific evidence, avoid misuses and misinterpretation, and explore ways to constructively use findings from cognitive neuroscience to improve the administration of justice." Through this initial investment, and the Foundation's renewed commitment with a grant of $4.84 million in 2011, the Law and Neuroscience Project has conducted extensive research, held conferences, created databases, and developed legal primers to help reconcile and accommodate various perspectives, supporting the responsible application of neuroscience to the increasing number of instances in which it has been introduced to support legal decision-making.

xviii Introduction

The metaphysical questions raised by the translation of neuroscience to criminal justice in relation to free will and determinism, and the associated sociopolitical debates regarding their relevancy to criminal responsibility, crime prevention, adjudication, and control, have been long-standing ones in the field of criminology.

Chapter 1 of this text provides a brief overview of the history of major criminology theories, and of the ways in which science, culture, and politics have influenced criminal justice policies and practices to date.

Understanding neuroscience's potential to advance our knowledge of the dynamics of criminal behavior requires foundational knowledge of the brain's structure and functioning, and of the technological advances that have rendered brain imaging possible. Chapters 2 and 3 provide an overview of the current state of knowledge regarding the regions of the brain, and details on the imaging systems that have accelerated efforts to understand its functioning.

Chapter 4 looks at the early attempts to apply brain science to the criminal justice system—both pre- and post-imaging—and how the use of neuroscientific evidence in several well-publicized cases may have influenced its rate of introduction and acceptance in the courts.

Chapter 5 looks inside the criminal brain, and details research on the brain–behavior correlates that have been identified to date.

Chapter 6 explores the application and limitations of neuroscience in the prevention and investigation of crime, including in relation to juvenile justice, leveraged treatment, victim memory, and lie detection.

In Chapter 7, this exploration continues with a review of the use of neuroscience during prosecution and sentencing.

Crime is a social construct and, consequently, approaches to crime and crime control are significantly influenced by culture. Chapter 8 reviews examples of applied neurocriminology in various nations.

An oft quoted 2002 article in *The Economist* on the ethics of what it called "brain science" warned, "Genetics may yet threaten privacy, kill autonomy, make society homogeneous and gut the concept of human nature. But neuroscience could do all of these things first."

Chapter 9 revisits the possibilities and the controversy posed by neuroscience, and the ramifications for contemporary and future approaches to crime and crime control.

The Las Vegas massacre, the recent year-over-year rise in violent crime in the United States, and increases in crime subtypes such as those perpetuated by "homegrown terrorists" and cyber criminals demonstrate the continuing and devastating public health and social challenges of crime and violence.

The following chapters explore the potential and limitations of neurocriminological solutions.

REFERENCES

Fink, S. (2017). Las Vegas Gunman's Brain Will Be Scrutinized for Clues to the Killing. *New York Times*.
Goodenough, O. R. (2001). Mapping cortical areas associated with legal reasoning and moral intuition. *Jurimetrics, 41*, 430–31.
Greene, J., & Cohen, J. (2004). For the law, neuroscience changes nothing and everything. *Philosophical Transactions of the Royal Society of London (Science B), 359*, 1775.
Montero, D. (2017). FBI Chief in Nevada Says Motive behind Las Vegas Concert Massacre Is Still a Mystery. *Los Angeles Times*.
Morse, S. J. (1996). Brain and Blame. Retrieved from http://scholarship.law.upenn.edu/faculty_scholarship/885.
People v. Weinstein 156 Misc.2d 34 (1992) 591 N.Y.S. 2d 715.
People v. Weinstein, 591 N.Y.S.2d 715, 718 (App. Div. 1992).
Raine, A. (2013). *The anatomy of violence: The biological roots of crime*. New York: New York Vintage Books.
Sapolsky, R.M. (2004). The frontal cortex and the criminial justice system. *Philosophical Transactions of the Royal Society of London (Science B), 359*, 1787.
Tavernise, S., Kovaleski, S. F, & Turkewitz. (2017). Who Was Stephen Paddock? The Mystery of a Nondescript "Numbers Guy." *New York Times*.
The Economist. (2002). The Ethics of Brain Science: Open Your Mind. May 23.
Transcript: Day One of the Roberts Hearings. (2005). Retrieved from http://www.washingtonpost.com/wpdyn/content/article/2005/09/13/AR2005091300693.
Williams, T. (2017). Violent Crime in U.S. Rises for Second Consecutive Year. *New York Times*.

Neurocriminology: How Did We Get Here?
A Brief History of Criminology

1

Crime is a fact of the human species, a fact of that species alone, but it is above all the secret aspect, impenetrable and hidden. Crime hides, and by far the most terrifying things are those which elude us.

Georges Bataille

Society prepares the crime; the criminal commits it.

Henry Thomas Buckle

Learning Objectives

1. Demonstrate knowledge of the major criminology theories.
2. Recognize the key differences between predominantly biologically based theories of crime and predominantly environmentally based theories of crime, and the impact on respective recommendations for criminal justice policies.
3. Analyze the effects of science, politics, and culture on criminology theories.

Introduction

Three questions related to criminality have vexed civilizations since the dawn of time:

1. Why do individuals engage in criminal behavior?
2. How can we prevent criminal behavior?
3. What do we do with those who commit crime?

Suggested answers have historically reflected a categorical alignment with one of two metaphysical perspectives: That we possess the will to act freely or, conversely, that our actions are predetermined by internal or external factors over which we have limited control. Politics, culture, and science have and continue to coalesce and collide, resulting in theories designed to determine which of these perspectives is more valid, reliable, and relevant to public policy.

Neurocriminology represents a 21st-century response.

It is a response informed by the successes and failures of the answers offered before it.

Whether one believes history is a series of pendulum swings over a relatively unchanging core, or an iterative evolutionary process that inevitably unfolds into the present, a review of some of the dominant criminological theories to date—those of the Pre-Classical Period, the Classical School, the Positivists, Mainstream Criminology, Critical Criminology, and Left Realist and Conservative Criminologies, as well as an example of the biologically based theories of the 21st century—illuminates the influence of politics, science, and culture on crime and crime control. The historic review also suggests the manner in which these factors have and will continue to influence neurocriminology.

The Pre-Classical Period: The Devil Made Him (or Her) Do It

The earliest theories of crime were intertwined with the spiritual and religious practices of the time. Individuals were believed to lack free will and thought to commit crime due to external forces, for example, the moon, animal spirits, and ancestors—each could cause the innocent to act contrary to accepted norms. Accordingly, punishment ranged from rituals designed to rid the violator of the corrupting influence or, in cases of particularly resistant forces, in isolation or exile. With the rise of organized religion, criminal behaviors came to be defined as violations of the customs of a particular church or sect; behaviors that countermanded adopted religious teachings were synonymous with violating the all-powerful will of a divine being. Such heretical acts were deemed the result of demonic possession, and the criminal was subject to various interventions ranging from prayer to imprisonment to death.

The 18th Century and the Classical School

In the 1700s, the will of man replaced the will of God, as the electorate replaced divine sovereignty. In this age of Enlightenment and reason, the Genevean philosopher Jean-Jacques Rousseau argued that legitimate authority can only be derived from a social contract agreed upon by all citizens for their mutual preservation, rather than through the edicts of a single leader presumed to be sanctioned by God.

The role of religious authority in politics was also challenged by Georges Cuvier's scientific confirmation of extinction, which contradicted the creationist view of life's inception and offered a biological basis for existence. Additionally, Thomas Bayes's theorem of probability—which suggests that

Neurocriminology: How Did We Get Here?

the likelihood that something that will occur can be statistically derived by looking at influencing factors—asserted that the universe is logical and malleable, rather than preordained and immutable.

Inventions of the time also reinforced the notion that fate is not predetermined. The period saw the development of the steam engine and the creation of the fabric spinning frame, heralding the beginning of the Industrial Revolution, in which manufacturing replaced manual labor and wider sections of the population had broader professional choices. Reflected in Adam Smith's publication of *The Wealth of Nations*, the century saw the rise of capitalism and, concurrently, a newly established and more independent middle class.

Within this context emerged the earliest recognized school of criminology: the Classical School.

Consistent with the rising rationalism within the political, scientific, and cultural sectors during this time, the Classical School proposed that crime was a logical consequence of an unfair system: criminals commit crime as a natural and understandable response to inequity. Preventing and controlling crime, therefore, would result from replacing arbitrary governance with laws that were fair and equitable, and through the establishment of legal processes that were clear and transparent.

Jeremy Bentham, one of the classical school's originators, epitomized the rational approach to crime and crime control. Bentham theorized that individuals were motivated by psychological hedonism; we all seek pleasure and avoid pain. Because of this, Bentham believed we were all capable of crime under the right circumstances. Our biologically driven urges were identical, and it was our social circumstances that dictate whether we decided to indulge in criminal actions.

From Bentham's perspective—and consistent with the scientific paradigm introduced by Bayes's theory of probability—an individual's actions could be predicted through a mathematical formula that Bentham labeled felicity calculus. Through this formula, Bentham postulated that he could calculate the specific amount of pleasure an individual would derive from engaging in a particular crime. The formula, in turn, could be used by government to determine the extent to which it needed to restrict civil liberties, an action inherently painful to the individual and, therefore, undesirable except as absolutely necessary to protect the majority. Anticipating debates that remain unresolved, Bentham also opposed criminalizing behaviors that offended individual morality but that arguably did not cause pain. During his time, for example, he opposed the punishment of vestal virgins, who were buried alive when found to be unchaste. Bentham argued that their behavior produced no true evil and, therefore, should not be subject to punishment. Similarly, Bentham proposed that punishment should only be as severe as necessary to serve as a deterrent. For this reason, Bentham was an early opponent of capital punishment (see Box 1.1).

BOX 1.1 BENTHAM'S PANOPTICON

Classical school criminologist Jeremy Bentham was a pioneer of efforts to rehabilitate criminals so they could successfully reenter society. Although considered then and now a bit bizarre, Bentham's panopticon is one of the earliest attempts to emphasis rehabilitation over retribution. His round structure featured prison cells with glass windows that were clearly visible to a guard, who was stationed in an enclosed center. The guard was not only responsible for watching the prisoners but also for serving as an employer, engaging inmates in paid manual labor from which the guard would receive a portion of the proceeds. Should a prisoner recidivate, the guard would be fined, thus supporting his investment in their rehabilitation. In Bentham's vision, panopticons would be located at the center of cities, as their high visibility would serve as deterrents to others tempted to engage in crime.

Bentham's panopticon idea was rejected in his native England, as well as in France and Ireland. However, two structures conforming to his design were attempted in the United States.

The Western State Penitentiary was opened in Pittsburgh in 1826 but deemed "wholly unsuited for anything but a fortress" and was ordered rebuilt in 1883.

A second attempt was made in Illinois between 1916 and 1924. As one inmate stated, "They figured they were smart building them that way" [with cells built in a circle and facing a center]. "They figured they could watch every inmate in the house with only one screw [warden] in the tower. What they didn't figure is that the cons know all the time where the screw is, too."

The building was abandoned and a more conventional structure replaced it.

Bentham's panopticon concept was reinvigorated in 1975 by French philosopher Michel Foucault. In his *Discipline and Punish*, Foucault used the panopticon as a metaphor for the way in which disciplinary societies subjugate citizens: "He is seen, but he does not see; he is an object of information, never a subject in communication." Consequently, the surveilled engages in self-policing for fear of punishment.

Bentham's original concept and Foucault's expanded metaphor have been adopted by some as warnings of the consequences of eroding privacy, ranging from the proliferation of cookies on Internet websites to the closed-circuit television monitoring of public spaces.

It will be interesting to see if advances in neurocriminology extend the application of the metaphor to this discipline as well.

Bentham's contemporary and fellow classical school theorist, the Italian criminologist Cesare Beccaria, likewise proposed that governmental authority should be restricted, and only extend to protecting individuals against war from enemies outside a nation's borders and the chaos that would otherwise reign within.

Beccaria was disturbed by the routine use of torture in Western Europe during his time, which served not only to punish but also as a means of determining innocence or guilt, compelling confessions, and identifying accomplices. In response, he wrote his famous work, *On Crime and Punishments.*

Beccaria argued that crime control would only be effective when individuals—who were by nature rational and capable of making rational choices—knew the punishment for particular criminal actions in advance. This, in turn, could only be accomplished by publishing laws and punishments, by rendering criminal proceedings public, and by ensuring that punishment was delivered shortly after a crime was committed.

The Classical Theorists offered a logical solution to crime and crime control that appealed to individual rationality, and was premised on one's ability to refrain from crime.

During the 19th century, several scientific and pseudoscientific developments would suggest that criminals did not have the volitional control upon which Classical Theorists relied.

The 19th Century: Phrenology, the Positivists, and Lombroso

In early 19th-century Germany, physician Franz Joseph Gall advanced a theory that would later be called phrenology. Based upon his observations of the sizes of the heads of gifted classmates, evaluation of the bumps on the skulls of pick pockets, and postmortem examinations of more than 100 brains, Gall posited that intellect and personality could be linked to configurations of the skull. He asserted that mental functions were localized in specific regions of the brain and could be measured by feeling the bumps, indentations, and overall shape of the individual's head. He mapped brain regions to brain functions, which he called faculties, including a faculty called "Combativeness" and one called "Acquisitiveness" (see Box 1.2).

His scientific basis for behavior was reinforced when naturalist and geologist Charles Darwin published *On The Origin of Species*, and later, *The Descent of Man: and Selection in Relation to Sex.* Darwin's theory of evolution, which challenged the notion that man originated from a single act of God, widened the science–religion divide that had emerged in the prior century. It also supported the perspective that human action is biologically driven.

BOX 1.2 PHRENOLOGY: EARLY EXPLORATIONS OF BRAIN REGIONS AND ASSOCIATED FUNCTIONALITY

Franz Gall's system of phrenology became an international sensation following its ban by last Holy Roman Emperor of the German Nation, Franz II, who feared that its materialistic emphasis violated moral and religious principles.

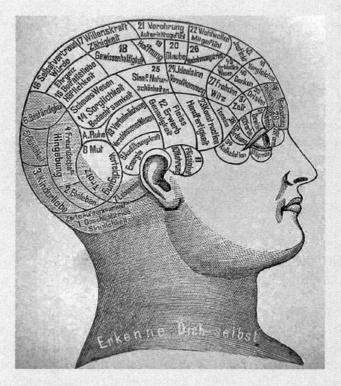

Source: Wikimedia Commons
Scan by de: Benutzer: Summi
Edited by "Wolfgang" [Public domain]

As reported by George Combe—lawyer and founder of the Edinburgh Phrenological Society—in 1835, Gall initially proposed 27 faculties associated with specific areas of the brain. These included the following:

1. Amativeness
2. Philoprogenitiveness
3. Adhesiveness

(Continued)

BOX 1.2 PHRENOLOGY: EARLY EXPLORATIONS OF BRAIN REGIONS AND ASSOCIATED FUNCTIONALITY (*Continued*)

4. Combativeness
5. Destructiveness
6. Secretiveness
7. Acquisitiveness
8. Self-Esteem
9. Love of Approbation
10. Cautiousness
11. Eventuality and Individuality
12. Locality
13. Form
14. Language
15. Language (included also in organ 14)
16. Coloring
17. Tune
18. Number
19. Constructiveness
20. Comparison
21. Causality
22. Wit
23. Ideality
24. Benevolence
25. Imitation
26. Veneration
27. Firmness

From the context of these and other developments, the Positivist School of Criminology was born. Positivists asserted that crime should be studied scientifically and that criminal actions cannot be limited to legal definitions, which are subject to the whims of societal sensibilities. The positivists also believed that the proclivity to engage in criminal activity was, to a large extent, predetermined. Consequently, because criminals had somewhat limited volitional control of their actions, the positivists advocated for the treatment of criminals, rather than punishment.

One of the most prominent names in the positivist school was Cesare Lombroso, who has been referred to as "the Father of Criminology."

Lombroso shifted the focus of criminology from criminal behavior—which had been a prominent focal point for his predecessors in the Classical School—to the criminal himself.

An army physician who oversaw an Italian insane asylum during his career, Lombroso authored the influential book *Criminal Man*, in which he argued that criminals could be identified by distinguishing physical characteristics, such as a sloping forehead, unusually sized ears, asymmetrical faces, and excessive arm length. Lombroso also asserted that the brains of criminals differed from noncriminal, a belief informed by an inspiration that he experienced when performing an autopsy on a notorious criminal in 1872: "I seemed to see all at once, standing out clearly illumined as in a vast plain under a flaming sky, the problem of the nature of the criminal, who reproduces in civilized times characteristics, not only of primitive savages, but of still lower types as far back as the carnivores."

This aspect of his criminal theory became known as Atavism, which referred to the reemergence of certain evolutionarily regressed characteristics in the contemporary criminal. Lombroso further postulated that physical characteristics allowed one to differentiate between certain categories of criminals, such as thieves, rapists, and murderers.

Lombroso took interest in the psychological characteristics of criminals as well and asserted that criminals are distinguishable by traits including lowered sensitivity to pain and touch, better eyesight, a lack of a moral compass, and greater vanity, impulsivity, vindictiveness, and cruelty. He stated that criminals are more prone to engage in excessive tattooing. These traits and predilections were seen by Lombroso as inherited rather than learned.

Given his belief that criminals were "born bad," Lombroso rejected the notion that punishment alone could prevent recidivism. Instead, Lombroso proposed that criminals should be rehabilitated through personalized scientific treatments based on the physical traits that rendered them vulnerable to criminal behavior. Presaging the specialty courts that exist today—such as drug and mental health courts—Lombroso proposed that treatments target a criminal's motivation.

Any unitary theory is challenged by its exceptions. Contemporaries of Lombroso offered different explanations as to why some individuals with the characteristics that Lombroso associated with criminality did not, in fact, commit crimes.

Enrico Ferri, foreshadowing arguments for a "compatibilist" conceptualization of free will, argued that the criminal was influenced by both biological and social factors, including population density, interactions with the justice system, and issues in the family of origin. Ferri also identified dynamic factors in the physical environment—such as temperature and the length of daylight—as influencing criminal activity.

Raffaele Garofalo, in contrast, rejected the notion of free will. For Garofalo, whether or not the biologically fated criminal engaged in crime

Neurocriminology: How Did We Get Here?

was determined by the degree to which he possessed "desirable" traits. Criminals who were "weakest" in terms of possessing these traits were predestined to commit crimes. As these individuals would not be able to resist engaging in criminal activity, the only solution to managing this criminal, Garafolo concluded, was to sentence him to death. In contrast, criminals who showed less weakness would commit crime only if offered the right opportunity. These criminals, Garofalo argued (in a manner that would be explored in the film *Minority Report* more than a century later), should either be sterilized to prevent breeding or be subject to long-term imprisonment. Finally, Garafolo allowed for a third category of criminal: those who have a predisposition toward crime but are unlikely to commit criminal acts unless desperate or forced to do so by circumstance. Garofalo recommended that this type of criminal be forced to make reparations to deter recidivism.

The 20th Century: Mainstream Criminology, Critical Criminology, Left Realism, and Conservative Criminology

Criminology in the 20th century saw the promulgation of a significant number of theories, reflective of seismic cultural, political, and scientific developments that were alternatively unifying and divisive.

During this period, the world was transformed by the World Wars I and II, followed by engagement in the Korean and Vietnam Wars. In the United States, the Industrial Revolution and capitalism of the prior centuries resulted in increased urbanization and affluence at the century's onset. The publication of Upton Sinclair's novel *The Jungle* and, less than two decades later, of F. Scott Fitzgerald's *The Great Gatsby* reflected different aspects of the appropriation of noble goals by the pursuit of wealth and pleasure. Each served as a cautionary tale regarding the excesses of the coveted American dream, proved prophetic by the onset of the Great Depression in 1929.

Several scientific advances joined the political and cultural backdrop that suggested reality is more nuanced that originally conceived. At the turn of the century, Max Planck originated the notion of quantum physics, which determined probable interactions on the atomic and subatomic levels. His discoveries supported the later invention of lasers and of the magnetic resonance imaging scanners used in much of contemporary neurocriminology research. Planck's work also inspired Albert Einstein's theory of relativity, which overturned Newtonian theories that time and space are fixed and immutable and, subsequently, introduced the then radical notion that individual perspective matters and influences one's perception of reality.

The 20th century also ushered in an information age that brought crime—particularly sensational crime—to public awareness as never before.

In this atmosphere, diverse criminological theories emerged, many of which reflected—or were reflective of—the increasing politicalization of crime and crime control. This was particularly apparent during the latter half of the 20th century, at which time the rate of violent crime quadrupled, and searches for both root causes and effective responses became a central focus of national and international dialogues.

The resulting 20th-century criminological theories can be broadly placed in four categories: mainstream criminology, advanced during the 1950s and early 1960s; critical criminology, which originated in the mid-1960s through mid-1970s; left realist and conservative criminologies, which were advanced in the 1980s and continue to influence perspectives of crime and punishment through the modern day; and early biologically based theories of the 21st century.

Mainstream Criminology

The first half of the 20th century—particularly following World War II—saw the rise of urbanization and a national focus on the pursuit of individual wealth and happiness. The stereotypical American family was reflected in popular TV series such as *Father Knows Best*, *I Love Lucy*, and *The Honeymooners*, while the idealized pursuit of justice was embodied in the character of Perry Mason. Yet several schools of criminology recognized that these cultural ideals eluded segments of population, and that criminal activity was disproportionately occurring in communities for which the American dream seemed less accessible. These criminological theories, collectively referred to as mainstream criminology, placed heavy emphasis on the criminogenic role of environment.

The first of these theories, known as the Chicago school of criminology, proposed various related theories. Two of the most prominent—Edwin Sutherland's Differential Association Theory and Ronald Akers' Social Learning Theory—shared the underlying premise that criminality was a learned behavior. Consequently, these theories asserted that crime control was best achieved by training criminals to more prosocial behavior by positive reinforcement and by removing them from environments that model antisocial behavior. Their theories led to intervention efforts such as the establishment of residential group homes that use "token economies" to reward rule following. One of the efforts inspired by mainstream criminology theories—the Chicago Area Project—continues to operate (see Box 1.3).

Neurocriminology: How Did We Get Here?

BOX 1.3   THE CHICAGO AREA PROJECT: AN EXAMPLE THE ENDURES

An enduring application of the Chicago school theorists' approach to crime control is the Chicago Area Project (CAP). CAP was founded by criminologist Clifford Shaw who, in concert with Henry D. McKay, applied social disorganization theory to understand and prevent juvenile delinquency. Specifically, the researchers built upon Ernest Burgess's concentric zone mapping theory that suggested levels of crime could be mapped to areas of ecological change that occurred around urban areas. Shaw and McKay found that delinquency rates were highest in what was referred to as the transition zone, that is, the urban area into which a central business district tended to expand, creating change and undermining existing social infrastructures. This, in turn, created three factors—poverty, ethic/racial heterogeneity, and population mobility—that were associated with social disorganization and subsequent delinquency.

CAP's approach includes creating neighborhood committees comprised of individuals from the community who encourage community youth to engage in prosocial behavior by modeling and through the introduction of recreational activities which serve as positive reinforcements for noncriminal behavior. CAP also includes efforts to improve the neighborhood's appearance, as well as training of community residents to engage in "curbside counseling."

Most studies of the efficacy of CAP have been criticized as flawed methodologically. A 1984, 50-year retrospective analysis published by the Rand Corporation, however, concluded the following: "All of our data consistently suggest that the CAP has long been effective in organizing local communities and reducing juvenile delinquency."

A second branch of mainstream theorists, known as anomie and strain theorists, suggested that criminality is fostered by overarching macro-level systems, rather than the micro-level ones of local communities. For these theorists, the structure and pressures of American society itself were criminogenic. The foundation for modern versions of anomie–strain theory were laid by Robert Merton, whose 1938 *Social Structure and Anomie* suggested that those who cannot attain economic success through legitimate means would pursue it through deviant acts. In contrast to the Chicago school theorists, strain theories suggested that it was the inability to leave the socioeconomically challenged communities—rather than the communities themselves—that was criminogenic.

Merton's strain theory received heightened attention during the tumultuous 1960s, as it resonated with the growing perspective that economic, gender, and cultural disparities were reinforced—if not encouraged—by social and political structures. Strain theory suggested that crime was a symptom of what many believed to be fundamental social injustices woven into the fabric of most societies.

As with most substantive issues, there were opponents to this view. Some researchers suggested that strain theory inappropriately assumed that all individuals were principally motivated by financial gain, a value that was not universally shared. Additionally, strain theory was criticized for its implication that those from lower economic classes necessarily felt greater strain and, therefore, were more disposed to engage in higher levels of deviant behavior, a perspective that ignored white collar and cybercrime, as well as cult violence and terrorism. Research has also found that those with high aspirations are more likely to conform and engage in prosocial behavior, a direct contradiction to strain theory's fundamental proposition that those with frustrated ambitions engage in criminal behavior.

Instead of dismissing strain theory outright, criminologist Robert Agnew suggested refinements to Merton's approach. To Agnew, status frustration represented but one kind of criminogenic strain. Agnew theorized that additional forms of strain contributed to criminality, including the loss of something valued—such as foreclosure on one's home—or negative treatment by others—such as being the victim of racial profiling. Agnew also addressed classic strain theory's limited ability to explain why, despite the experience of strain, some individuals committed crimes while others did not. Agnew proposed that differences were anchored in any number of factors, including an individual's ability to cope with strain, or to substitute one set of goals for more attainable ones, or to exercise self-control. Agnew also proposed that individuals with prior criminal histories or learned criminal behaviors were more vulnerable to respond in criminal ways. Finally, Agnew postulated that those who generally blamed others for their negative experiences—what psychologists call having an external locus of control—were more likely to respond to strain with criminality.

If the fundamental unfairness of society was the root cause of criminality, as was postulated by the strain/anomie theorists, then crime could only be controlled by leveling the playing field. Proponents of these theories pointed to the need to ensure that all had access to equal opportunities for success. Ensuring an urban school in a low-income neighborhood was equipped with computers for its students, for example, was not sufficient for strain theorists. Instead, these students needed access to the same level of teaching talent that more affluent schools enjoyed, a perspective typically embraced by those who held liberal political views. In contrast, some strain theorists argued that the way to counter crime was to strengthen individual ties to noneconomic

Neurocriminology: How Did We Get Here?

institutions, such as family systems and places of worship, a perspective embraced by many political conservatives.

Concurrent with strain/anomie theories, an additional theoretical perspective was advanced. Known generally as control theories, this theoretical perspective suggested that answering "Why individuals commit crime?" was not as relevant as answering "Why individuals do not?"

Control Theories

Early control criminologists, including Albert Reiss and F. Ivan Nye, proposed that criminality resulted when institutions, whether at the familial, community, or societal level, were unable to command obedience to social norms. Ineffective institutional control was evidenced when the needs of members were not met, or when nonconformity was inadequately sanctioned.

An additional control theory, which gained great influence at the time of publication, was Walter Reckless's containment theory. Consistent with other control theorists, Reckless suggested that lack of engagement in criminality was a natural byproduct of the external containment provided by organized groups such as families, schools, and communities that reinforced conventional behavior, encouraged internalization of rules, and offered supportive relationships. This explained why crime was higher in socially disorganized communities. However, even in such communities, the majority of individuals did not commit crimes. Reckless explained the lack of criminality among those in socially disorganized communities as resulting from a second type of containment—inner containment—which allowed individuals to resist pushes toward crime (such as poverty) and pulls toward crime (such as offending peers). Inner containment, what psychologists label resiliency, included a positive self-concept that valued being a law-abiding citizen, appropriate goal orientation, frustration tolerance, and the ability to effectively cope with failure or disappointment. Key to Reckless's theory was its refutation that social structure—whether on the societal or community level—determined engagement in crime, a premise that Reckless considered overly deterministic.

Control theorists Gresham Sykes and David Matza agreed with Reckless's assertion that many criminological theories of the time overemphasized the influence of the environment on criminal behavior. Sykes and Matza sought to explain why, if social influences were so powerfully destructive, delinquent and nondelinquents alike were fairly conventional in many ways. The youth they studied only engaged in criminal acts when they intentionally rejected the controls of conventional society, explaining why the majority only engaged in crime episodically rather than persistently. Sykes and Matza contended that the socialization process itself taught youth ways to reject conventional norms through what they referred to as techniques of neutralization. These techniques included denial of responsibility ("I didn't mean to do that"), denial of injury ("I didn't really hurt anybody"), denial of victim ("They only got what they deserved"), condemnation of the condemners ("Everyone is being unfair in judging me"), or appeal to higher loyalties ("I did it for the good of my family, not for myself").

Travis Hirschi's initial theory of social control considered the intersection of the individual and the environment. Hirschi began with an underlying assumption that all individuals could see the benefit of committing crimes and, therefore, the motivation to engage in criminal activity was universally shared. What varied, according to Hirschi, was the strength of an individual's social bonds, which in turn, determined the extent of social control upon his or her behavior. Hirschi identified four such bonds: Attachment, Commitment, Involvement, and Belief. Attachment represented emotional attachment to others, particularly parents, who controlled behavior directly by rules and indirectly by setting expectations that the child did not want to disappoint. Commitment represented high educational and occupational aspirations, which reinforced law-abiding behavior as these goals could be derailed by criminality. Involvement was defined as participation in activities such schoolwork, sports, and other recreational activities, which left little time to offend. Belief was an embrace of the validity of rules and laws. Compromised bonds, which were not *a priori*, could lead to criminality.

A little more than two decades following the publication of his Social Bond Theory, Hirschi advanced a very different perspective when he joined with Michael Gottfredson and authored *A General Theory of Crime*. In this work, the theorists proposed that self-control, rather Hirschi's earlier assertion of social control, determines criminal involvement. Self-control, the theorists asserted, emerged at a young age and was the ability to resist engagement in behaviors that the average person, upon reflection, would find criminal or even dangerous or foolish. Those with low self-control were apt to take unnecessary risks in all aspects of their lives, including when driving, in interpersonal relationships, and in the workplace. What transformed these individuals into criminals was the opportunity to commit crime.

For Hirschi and Gottfredson, low self-control was humanity's natural state; we were all born with low self-control. Only through strong socialization provided by effective parenting did we gain the self-control that would prevent risk taking and criminality. Strong parenting was characterized by the parents' ability to effectively monitor their child, to recognize when the child was obedient, and to punish or correct any deviant behavior.

The control theories of Reckless, Sykes, Matza, and their 1950s contemporaries, which were echoed in the later theories of Hirschi, emphasized the importance of conventional institutions and conformity in preventing crime. Their theories supported the development of crime control models that were primarily preventive; only by intervening early would individuals neutralize the contrary values and impulses that led to criminal engagement.

Neurocriminology: How Did We Get Here?

The emphasis upon fostering conformity was consistent with the sociopolitical culture of the time. Americans responded to the tumultuousness of events such as the Great Depression and World War II by embracing conventional norms and behaviors.

In the 1960s, the valued conformity that had been experienced as a breath of fresh air began to suffocate broad sections of the population.

During this time, the successors to control criminologists—the critical criminologists—suggested a paradigm shift in attempts to understand criminality. For these theorists, maintaining a fundamental skepticism—if not an outright rejection—of the unconditional acceptance of social norms was critical to addressing the issue of crime.

Critical criminologists asserted that social acceptability was narrowly constructed and generally reflective of middle-class values that were not shared by all. The imposition of these values on those outside the middle-class structure determined what was considered criminal and, equally important, what was not. Deviance, then, was relative and reinforced by social structures that refused to acknowledge the validity of other perspectives.

Although there are many important criminological theories that fall, broadly, under the umbrella of critical criminology (e.g., shaming theory, conflict theory, and feminist theory), one that exemplifies the underlying principles is labeling theory.

For labeling theorists, primary deviance was fairly common and results from a variety of environmental and psychological factors. Most often, an individual's response to engagement in initial deviance was to decide that the behavior was out of character. Recidivism results from societal reaction to the deviant act; if the offender was stigmatized (e.g., labeled a criminal), the individual would begin to adopt the persona of the criminal, and incorporate this persona into his or her self-concept. The stronger the societal reaction, the more likely this dynamic would occur, and the more certain the individual would behave in accordance with this view. Labeling, then, became a self-fulfilling prophecy.

Labeling theorists proved their theory through a number of studies conducted in the 1960s and 1970s. In one, researchers randomly placed Black Panther bumper stickers on the cars of a racially mixed group of Los Angeles college students, each of whom had exemplary driving records. Within hours, those with the bumper stickers accumulated numerous driving citations, such as for improper lane changes. Similar results were found in a study of two groups of high school males, one from middle-class environments and one from less affluent families. The groups showed disparities in dress and social skills. However, despite similarities in quantity and quality of deviant behavior, the students from the lower socioeconomic group were more often the subject of stops by law enforcement than those from the middle-class families.

The premise that individuals become criminals as a result of being labeled as such caused labeling theorists to argue that the criminal justice system itself was criminogenic. They suggested that once someone was adjudicated as a criminal, the odds of recidivism increased. Consequently, labeling theorists promoted four different paths to controlling crime. The first involved the decriminalization of what they deemed to be victimless crimes, such as public drunkenness, drug use, gambling, and pornography. The second was diversion from the criminal justice system for minor crimes such as truancy. Third, the labeling theorists supported the institution of due process rights then not guaranteed, such as the accused's right to an attorney and protection against unwarranted searches. The fourth approach to crime control suggested by labeling theorists was deinstitutionalization, achieved by halting the construction of prisons, and releasing certain classes of offenders to community-based programs.

Although all of these recommendations would be subsequently adopted to certain degrees and with varied levels of success in the intervening decades, the underlying hypothesis of labeling theory and other critical criminology theories—that the basic structure of society is unjust and criminogenic—failed to provide an actual blueprint for policy that could be realistically and practically implemented to address criminality and control crime. Concurrently, the pendulum of politics swung toward a more conservative worldview following the upheavals of the 1960s and 1970s. The longing for simplicity significantly influenced public policy initiatives, including those related to crime control. Criminal justice policy reverted to a straightforward "if, then" approach: If an individual committed crime, then he or she should be punished. Left Realism emerged as a way to counter this trend. It sought to address the failings of critical criminology by creating practical crime policy, while fighting against what it perceived to be an overly reductionistic "one punishment fits all" approach to crime that began to emerge in the late 1970s and early 1980s.

Left realism has its roots in Britain and was advanced by a group of criminologists including Jock Young. Young, who is credited with expanding interest in criminology in higher educational institutions throughout the United Kingdom, also served on the faculty of John Jay College and the City University of New York Graduate Center.

A central feature of left realism was an effort to describe crime and suggest crime control solutions that were not based solely on deterministic notions, such as impoverishment or genetic disposition. Instead, left realists posited that crime was the result of several factors including relative deprivation—the experience that one was less fortunate than those in similar groups—which left realists suggested was particularly prevalent in the 20th century, during which economic and social disparities were rising. Left realists also suggested that crime resulted from marginalization; when groups

Neurocriminology: How Did We Get Here?

lacked representation of their interests, and lacked clearly defined goals, violence was more likely to be employed as a form of political action.

An additional facet of left realism was inclusion of the victim, the public, and law enforcement in explanations of crime. Specifically, left realists suggested that crime could only be understood by also considering the factors that rendered individuals vulnerable to victimization, and by considering public attitudes and responses to crime and crime control practices, particularly those employed by law enforcement.

Left realism's multifaceted approach to understanding crime was codified into a theoretical system known as the Square of Crime, which represented four interrelated elements to understanding criminality: The state and its agencies; the offender and his or her actions; informal methods of social control (sometimes called "society" or the "public"); and the victim. Understanding the inter-relationships between these four elements, according to left realists, was key to understanding criminal activity, and crime control could only be successful if informed by a valid theory of crime. Absent scientific grounding, left realists believed that law enforcement would drift toward what they termed "military policing" police that would include stopping and frisking categories of people determined to be at possible risk for committing crimes, or using surveillance of large groups, including the innocent, to identify suspects. This, left realists asserted, would contribute to rather than prevent violence by mobilizing bystanders and creating unrest. Additionally, such preemptive policies were typically dismissed by left realists as distractions from the real solution, which involved catalyzing communities to engage in crime prevention.

To this end, left realists called for improvements in police–community relationships, which would support a flow of information that allowed for more precise targeting of crime and criminals. Finally, left realists advocated for the use of alternatives to prison in dealing with criminals, contending that, on a practical level, prison did little to rehabilitate—and more to harden—criminals who would return to society, resulting in increased rather than decreased crime rates.

The sociopolitical context that bred left realism criminology had an even greater impact on the advancement of conservative criminology, particularly in the United States. President Jimmy Carter closed the 1970s with a speech that focused upon America's "crisis of confidence." In the time immediately preceding his speech, OPEC increased oil prices, creating gas shortages that resulted in long lines and short tempers. This was viewed as another in a list of insults to American idealism and superiority that included the fall of South Vietnam, the Watergate Scandal, an accelerated international arms race, and high unemployment, inflation and interest rates. In his speech, Carter suggested that, "The solution of our energy crisis can also help us to conquer the crisis of the spirit in our country." He added an admonishment, "In a nation that was proud of hard work, strong families, close-knit communities and our faith in God, too many of us now tend to worship self-indulgence and consumption. Human identity is no longer defined by what one does but by what one owns."

The initially well-received speech was immediately met by a backlash, as Americans objected to being blamed for the nation's challenges. This resentment, coupled with a failed rescue of 52 American diplomats and staff held hostage at the U.S. Embassy in Iran, contributed to the Democratic president's resounding defeat by Ronald Reagan.

In contrast to Carter's pessimism and caution, Reagan's platform was based on confidence in American's potential and future, a confidence that appeared to be well-placed when the American hostages in Iran were formally released to U.S. custody minutes after Reagan was sworn into office. Reagan also appealed to those who sought more direct solutions to social challenges, including those of crime. This was exemplified in remarks he gave at the 1984 Annual Conference of the National Sheriff's Association: "It is interesting, too, to note that common sense about crime is making its impact in the very field which once accounted for so much of the misguided advice about crime, that of the social sciences." He went on to state that "Choosing a career in crime is not the result of poverty or an unhappy childhood or of a misunderstood adolescence. It's the result of a conscious, willful, selfish choice made by some who consider themselves above the law, who seek to exploit the hard work and, sometimes, the very lives of their fellow citizens."

Reagan's conservative social stand is consistent with the fundamental principle of the various theories that fall under the umbrella of conservative criminology: Crime was not the result of root causes that could be addressed by government. Rather, crime was a choice made of free will, and those who committed it needed to be met with swift and decisive punishment, which also served as its deterrent.

This perspective is represented in several conservative criminologies, including those advanced by James Wilson, Richard Herrnstein, Charles Murray, Stanton Samenow, Morgan Reynolds, William Bennett, John DiIulio, and Ronald Walters.

In their 1985 book, *Crime and Human Nature*, Wilson and Herrnstein—echoing the earlier work of Lombroso—suggested that criminals have "constitutional factors" that rendered them more inclined to crime, including distinctive body types. Additionally, they made the claim that "bad families produce bad children," a claim they buttressed by identical twin studies that showed those separated at birth and raised in different environments were nonetheless likely to have committed crimes.

Teaming up with researcher Charles Murray, Herrnstein proposed a slightly more specific explanation for criminality, and suggested that

Neurocriminology: How Did We Get Here?

intelligence—particularly IQ—is a significant constitutional factor that contributed to criminality. The theorists addressed this factor in social context, and contended that in a postindustrial world, with an increased emphasis on knowledge and advanced technical expertise, it was more difficult for those who were cognitively disadvantaged to compete, resulting in a lifetime of failures in relation to engagement in school, the workplace, their families, and communities. This, in turn, created a greater likelihood of engagement in criminal activities.

Stanton Samenow, whose work influenced Reagan's remarks to the National Sheriff's Association, suggested that criminals were pathological. Specifically, Samenow posited that criminals thought differently than noncriminals, that the criminal orientation was egocentric and grandiose, and that they were impulsive, easy to anger, and lacked empathy. Additionally, Samenow theorized that criminals externalized blame and manipulated others for self-satisfaction. Samenow's definition of the criminal mirrored the psychological construct of the psychopath, which is estimated to represent between 10% and 25% of the incarcerated population. Still, Samenow posited that all criminals shared these traits and that they self-selected into a criminal life.

Economist Morgan Reynolds also suggested that criminals are self-selected. Unlike Samenow, Reynolds asserted that criminality was a choice based upon a calculation of expected benefits versus expected costs; the prevalence rates for crime results from the benefits outweighing typical punishment, rendering crime "more attractive than other career options."

In 1996, William Bennett, former Secretary of Education under Reagan and former Director of the Office of National Drug Control Policy under George H.W. Bush, joined with John DiIulio, and Ronald Walters to suggest a slightly different theory of crime. Specifically, the three posited that crime resulted not from economic poverty but from moral poverty. Moral poverty occurred when children lacked loving, supportive parents who habituated them into prosocial decision-making.

Although the conservative theorists disagreed on the specific causes of criminality, they concurred on the appropriate response: Each suggested longer, swifter, and more straightforward punishment—first at the parental level and, should this fail, at the governmental level. Such crime control measures were consistent with the underlying philosophy that, whether the result of IQ, personality, choice, or moral poverty, criminals would continue to engage in crime unless the choice to do so was rendered unappealing.

Both left realism and conservative theories of criminology continue to influence much of modern criminal justice policy. The turn of the century also saw several additional events and developments that have had strong reverberations through the subsequent decades.

20   Neurocriminology

The 21st Century: Precursors to Neurocriminology

In 2000, the highly contentious U.S. election was ultimately decided by the Supreme Court which, in *Bush v. Gore*, halted a contested recount of Florida ballots, resulting in George W. Bush's win of the electoral college and the White House. The narrow and indecisive victory further polarized a nation that simultaneously expressed a heightened distrust of government and a willingness to view the world through a prism of partisanship that was supported by the rival liberal and conservative cable news programs founded at the end of the 20th century. In the ensuing decades, these media would echo division on every major issue, including crime and punishment.

A second event at the onset of the 21st century would initially unite and subsequently further divide the nation and the world. On September 11, 2001, Al-Qaeda terrorists hijacked four commercial airliners, crashing two into the World Trade Center, one in a field in Pennsylvania, and one into the Pentagon, killing nearly 3,500.

Additional acts of terrorism at home and abroad focused national attention on this form of crime. Cybercrimes, elder abuse, as well as significant school, workplace, theatre, and officer-involved shootings also rendered crime control a continued, urgent, and unresolved public policy concern.

Concurrently, criminologists grappled with the perplexing reality that economic disparity, social injustice, and unemployment were not as directly correlated with violent crime as originally believed. As social conditions in the United States deteriorated under the great recession, crime rates did not rise; FBI figures showed that murder, rape, robberies and other serious crimes fell to their lowest levels in nearly half a century.

Eroding social explanations for the etiology of criminality coincided with a myriad of scientific advances that shifted emphasis back toward a biological basis of crime. In contrast to times past, however, modern scientifically based theories incorporate psychosocial criminogenic factors, offering more nuanced—and more complex—explanations of crime. (Some explanations have met with decidedly mixed results. See Box 1.4, for example.)

In 2000, for example, researchers Lee Ellis and Anthony Walsh offered an explanation of crime that contextualized its biological bases within the environment. They began with the relatively straightforward observation that some individuals required greater levels of environmental stimulation than others and, consequently, engaged in different types of behaviors. In studying this phenomenon more closely, the researchers linked arousal states to the reticular activating system (RAS), the brain cells located at top of the spinal cord that are responsible for mediating environmental stimuli. They discovered that, while the majority of individuals experienced an optimal level of stimulation when environmental conditions strike a balance between being too constant and too varied, some individuals possess an RAS that is

Neurocriminology: How Did We Get Here?

highly sensitive to stimulation, and others an RAS that is unusually insensitive to stimulation. The researchers labeled the former group of individuals "augmenters" and the latter "reducers." Reducers tended to have a correlated hypoactive autonomic nervous systems (ANS), the part of the nervous system responsible for control of bodily functions not consciously directed, including breathing, heartbeat, and digestive processes. ANS is also the part of the nervous system responsible for involuntary reactions to dangerous situations, that is, the "fight or flight response."

Reducers experienced underarousal of the ANS—associated with fearlessness—and underarousal of the RAS—associated with sensation seeking. These individual were seek out environments in which they could participate in higher-risk activities, during which they would experience less fear than most, in order to attain a state of emotional equilibrium.

Ellis and Walsh's suboptimal arousal theory was consistent with numerous studies that employed electroencephalography (EEG) machines to assess the brainwave patterns and resting heart rates of criminals; relative to the general population, those who experienced underarousal were found to be more likely to engage in criminal activity.

Suboptimal arousal theory became one of the first explanations for the association between brain functioning and criminality to emerge in the current century. As will be explored in the following chapters, the translation of neuroscience to crime has offered significant additional insight into the brain–behavior–personality correlates of criminals.

BOX 1.4 WARRIOR GENES AND MISINTERPRETATIONS

In the early 1990s, the reported a link between violence and a gene on the X chromosome made international news following *Discover* magazine's publication of research related to a particularly large and unusual Dutch family. The male members of the family had a multi-generational history of extreme violence that dated to the 1870s. Two of the men were arsonists. One attempted to rape his sister. One tried to run over an employer with a car. One attempted to assault a mental health hospital staff member with a pitchfork. These men also demonstrated intellectual disabilities, with reported average IQs of 85.

Geneticist Hans Brunner and his colleagues discovered a correlation between the violent males and a gene on the X chromosome that encodes for monoamine oxidase A (MAOA). MAOA is an enzyme that regulates neurotransmitters including dopamine, serotonin, and norepinephrine, which are associated with behavioral and emotional

(Continued)

BOX 1.4 WARRIOR GENES AND MISINTERPRETATIONS (*Continued*)

regulation. (Additional information regarding the role of neurotransmitters is provided in Chapter 2.)

Brunner cautioned that the there was no such thing as "an aggressive gene," and that all behavior needs to be viewed as an interplay between multiple influences, including environmental factors. Nonetheless, the concept that certain individuals are "born bad" caught the media and the public's imagination.

Later, researchers reported a correlation between violent aggression and MAOA-L, a variant of the MAOA gene. The finding was particularly intriguing as the MAOA variant is also found in apes, which led to speculation that it conferred an evolutionary advantage.

In 2004, *Science* magazine dubbed MAOA-L "the warrior gene."

In 2006, researchers in New Zealand reported that MAOA-L occurs in 56% of Māori men, a Polynesian people indigenous to the county. The goal of the research, which also found that MAOA-L was found in 34% of Caucasians, 29% of Hispanics, 59% of Africans, and 77% of Chinese, was to determine if monoamine oxidase (MAO) is a genetic marker for alcohol and tobacco dependence. In an interview following presentation of their findings, however, the researchers were quoted as stating, "It is well recognized that, historically, Māori were fearless warriors."

The statement once again prompted speculation that there is a genetic basis for aggression. It also raised questions regarding the implications of the warrior gene and racial profiling.

Geneticists and bioethicists were quick to correct the sensationalism that ensued. An article published the following year stated:

"Whilst there is credible evidence for a contribution of a monoamine oxidase-A genetic variant to antisocial behaviour in Caucasians, there is no direct evidence to support such an association in Māori. Insufficient rigour in interpreting and applying the relevant literature, and in generating new data, has (in conjunction with a lack of scientific investigative journalism) done science and Māori a disservice."

The original researchers also worked to quell the controversy through a series of statements and interviews that included a 2015 publication that incorporated the following: "However, much of the controversy was unjustified because it stemmed from a combination of misquotes and misunderstandings printed in the original article released by the Australian Press Association."

(Continued)

Neurocriminology: How Did We Get Here?

BOX 1.4 WARRIOR GENES AND MISINTERPRETATIONS (*Continued*)

Despite the retractions and clarifications, the legend of the warrior gene continued to grow.

In 2008, researchers determined that those with the gene were more likely to punish others (defined as "administering varying amounts of unpleasantly hot [spicy] sauce to their opponent") when provoked. And in 2009, additional research suggested that males with MAOA-L were more likely to report being gang members, although the study also found that approximately 40% of the gang members in the sample did not carry the gene variant.

The sensationalism and misinterpretation of the warrior gene data is perhaps an understandable response to society's quest for a definitive etiology of violence. It is also a cautionary tale for scientists seeking to inform this search.

As Professors Peter Crampton and Chris Parkin wrote following the Māori study of 2006: "We conclude that in all science, and particularly where there is a highly charged social and political setting, the scientist has a responsibility for the way in which findings are disseminated and for ensuring a clear public understanding of the limitations of the work."

Key Terms

Pre-Classical Criminology: Early criminology theories that assert that individuals lack free will and that any deviation from accepted social or religious norms is caused by supernatural forces.

Classical School of Criminology: Theories the emerged during the Enlightenment period that posit that individuals chose to commit crimes, principally motivated by the desire to seek pleasure and avoid pain. Major theorists include Jeremy Bentham and Cesare Beccaria.

Positivists School of Criminology: Theories originating in the 19th century that focus upon the criminal, rather than the crime, and that assert that criminals should be studied scientifically. Major theorists include Cesare Lombroso, Enrico Ferri, and Raffaele Garofalo.

Mainstream Criminology: A collection of criminological theories that originated in the 20th century that place heavy emphasis on the criminogenic role of environment. These theories include Chicago school of criminology theorists Edwin Sutherland and Ronald Akers' Social Learning Theory, and anomie and strain theorists Robert Merton and Robert Agnew.

Control Theories of Crime: A collection of criminological theories originating in the 20th century that focus upon why individuals in similar circumstances do not become criminals. Major theorists include Albert Reiss, F. Ivan Nye, Walter Reckless, Gresham Sykes, David Matza, Travis Hirschi, and Michael Gottfredson.

Critical Criminology: A collection of criminological theories originating in the 20th century that assert that societal norms are narrowly constructed, and deviance results from an imposition and reinforcement of arbitrary values. Theories include shaming theory, conflict theory, feminist theory, and labeling theory.

Left Realist Criminology: Criminological theories of the latter half of the 20th century that suggest that crime is the result of several factors including relative deprivation, marginalization, and an interplay of the state and its agencies, the offender and his or her actions, informal methods of social control, and the victim.

Conservative Criminology: Criminological theories of the latter half of the 20th century that suggest that crime is not the result of root causes that can be addressed by government, is influenced by a variety of factors, and is ultimately the choice of the criminal.

Use Your Brain

Test Your Knowledge

1. In early 19th century, Franz Joseph Gall advanced a theory that would later be called phrenology. Based upon his observations, Gall posited that:
   a. Criminals have head sizes that are significantly larger than noncriminals.
   b. Criminals have head sizes that are significantly smaller than noncriminals.
   c. Mental functions were localized in specific regions of the brain.
   d. Criminals had larger foreheads and, therefore, less capacity for mental functioning.
2. Atavism is:
   a. Associated with Cesare Lombroso and refers to the reemergence of certain evolutionarily regressed characteristics in the contemporary criminal.

Neurocriminology: How Did We Get Here?

    b. Associated with Franz Gall and refers to the ridges on a criminal's skull corresponding with proclivity for criminality.

    c. Associated with Jeremy Bentham and refers to the approach to designing prisons.

    d. Associated with Raffaele Garofalo and refers to the approach to treating criminals.

3. One element of labeling theory asserts that:

    a. Criminality is an arbitrary social construct.

    b. Using inflammatory characterizations for certain crimes is, in itself, criminogenic.

    c. Societal reaction to a deviant act can create a self-fulfilling prophecy.

    d. If crimes were labeled by numbers, instead of with names, crime rates would decline.

4. During the 21st century, one challenge faced by criminologists has been to explain the disconnect between challenging social conditions and crime rates.

    a. True

    b. False

5. According to suboptimal arousal theory, a risk factor for criminality includes:

    a. A hyperactive autonomic nervous system and a hypoactive reticular activating system.

    b. A hypoactive autonomic nervous system and a hypoactive reticular activating system.

    c. A hyperactive autonomic nervous system and a hyperactive reticular activating system.

    d. A hypoactive autonomic nervous system and a hypoactive reticular activating system.

Apply Your Knowledge

1. Across the centuries, criminological theories have tended to emphasis the role of the environment or the role of biology in explanations of criminal behavior. Should etiology matter to society's response to crime?

2. Science, culture, and politics have consistently played a role in criminal justice practices. Given this inevitability, how might these influences be better harnessed to inform public policy?

Answer Key:

1. (c) 2. (a) 3. (c) 4. (a) 5. (b)

Bibliography

Andrews, D., & Bonta, J. (2015). *The psychology of criminal conduct.* New York: Routledge.

Beccaria, C. (1963). *On crime and punishments.* New York: Pearson.

Bhugra, D. (1996). *Psychiatry and religion: Context, consensus and controversies.* New York: Routledge.

Blackstone, W. (1753). *Commentaries on the laws of England in four books, vol. 1.* The Online Law of Liberty. Retrieved from http://oll.libertyfund.org.

Bosworth, M. (2005). *Encyclopedia of prisons and correctional facilities.* Thousand Oaks, CA: Sage Publications.

Buckle, H. T. Retrieved from http://web.inter.nl.net/hcc/rekius/buckle.htm.

Carter, J. (1979). *Address to the Nation on Energy and National Goals: "The Malaise Speech."* The American Presidency Project. Retrieved from www.presidency.ucsb.edu/ws/?pid=32596.

Combe, G. (1853). *A System of Phrenology* (5th ed.). London: Forgotten Books.

Crampton, P., & Parkin C. (2007). Warrior genes and risk-taking science. *New Zealand Medical Journal, 120*(1250), U2439.

Dale, A. (2003). *Most honourable remembrance: The life and work of Thomas Bayes: sources and studies in the history of mathematics and physical sciences.* New York: Springer-Verlag.

Darwin, C. (2003). *On the origin of species.* New York: Penguin Publishing Group.

Darwin, C. (2005). *The decent of man: And selection in relation to sex.* London: Forgotten Books.

Gamow, G. (1966). *Thirty years that shook physics: The story of quantum theory.* New York: Doubleday & Co.

Geis, G. (1955). Pioneers in Criminology VII—Jeremy Bentham (1748–1832). *The Journal of Criminal Law, Criminology and Police Science, 46*: 159–171.

Gibbons, A. (2004). Tracking the evolutionary history of a "Warrior" Gene. *Science, 304*(5672), 818. doi:10.1126/science.304.5672.

Gribbon, J. (2004). *The scientists: A history of science told through the lives of its greatest inventors.* New York: Random House.

Fitzgerarld, F. S. (1953). *The great Gatsby.* New York: Scribner.

Jeffery, C. (1959). The historical development of criminology. *Journal of Criminal Law and Criminology, 50*: 3–19.

Lilly, J., Cullen, F., & Ball, R. (2015). *Criminological theory: Context and consequences* (6th ed.). Los Angeles, CA: Sage Publications.

Lombroso, C. (2006). *Criminal man.* London: Duke University Press.

Merriman, T., & Cameron, V. (2007). Risk taking: Behind the warrior gene story. *New Zealand Medical Journal, 120*(1250), U2440.

Mitchell, A., & Kemp, J. (Eds.). (2009). *The obsessions of Georges Bataille: Community and communication.* New York: State University of New York Press Albany.

Reagan, R. (1984). *Remarks at the Annual Conference of the National Sheriff's Association in Hartford, Connecticut.* The American Presidency Project. Retrieved from www.presidency.ucsb.edu/ws/index.php?pid=40074.

Reynolds, M. (1985). *Crime by choice: An economic analysis.* Dallas, TX: Fisher Institute Publications.

Richardson, S. (1993). A Violence in the Blood. *Discover Magazine.* Oct. 1.

Neurocriminology: How Did We Get Here?

Riedel, M., & Welsh, W. (2011). *Criminal violence: Patterns, causes and prevention* (3rd ed.). New York: Oxford University Press.

Rudwick, M. (1997). *George Cuvier, fossil bones, and geological catastrophies.* Chicago, IL: University of Chicago Press.

Schlossman, S., Zellman, G., & Shavelson, R. (1984). *Delinquency prevention in South Chicago: A 50-year perspective of the Chicago Area Project.* Santa Monica, CA: Rand Corporation.

Sinclair, U. (2014). *The Jungle.* Tampa, FL: Millennium Publications.

Worrall, J. (2015). *Crime control in America: What works?* (3rd ed.). Upper Saddle River, NJ: Pearson.

Brain Basics
How Neurocriminology Is Possible

2

The human brain has 100 billion neurons, each neuron connected to 10 thousand other neurons. Sitting on your shoulders is the most complicated object in the known universe.

Michio Kaku

The chief function of the body is to carry the brain around.

Thomas A. Edison

Learning Objectives

1. Describe the role of neurotransmitters in brain functioning.
2. Identify the regions of the brain and the functions generally correlated with these regions.
3. Recognize the common causes of brain dysfunction.

Introduction

Three pounds is responsible for determining if we breathe and how we move, the manner in which we communicate and respond, what we think and how we feel, how we consolidate, process, and remember information, and how we make this information meaningful. Three pounds is responsible for who we are.

For centuries, philosophers, psychologists, and physicians have recognized that these three pounds—the brain—control mental processes related to sensation, thought, and memory. Seminal cases, such as those of Phineas Gage and H.M., remain standard in the curriculum of students of psychology, biology, and neuroscience (see Box 2.1). Yet it is only in the last several decades that science has begun to identify and distinguish the fundamental mechanics of brain processes.

A basic understanding of the structure and functioning of the brain, and awareness of the common causes of brain dysfunction, provide the foundation for understanding neurocriminology.

BOX 2.1 THE SEMINAL CASES OF PHINEAS GAGE AND H.M.

Two of the most enduring cases exploring the inferential relationship between brain functioning and behavior are those of Phineas Gage and H.M.

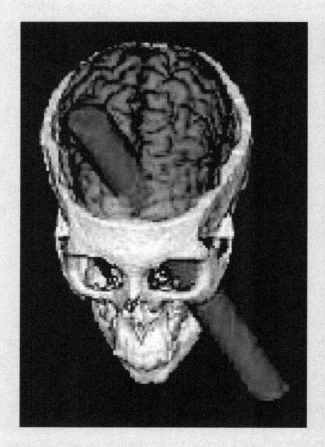

(https://encrypted-tbn0.gstatic.com/images?q=tbn:ANd9GcSQV-AoMOICoZJdWuZAX30tVIErhjcoUpQOm4HXxlTURJJsowpp2w)

PHINEAS GAGE

In 1848, the 25-year-old Gage was managing a railroad construction crew in Vermont when the tamping iron he used to remove boulders from the railroad's path caused a spark that prematurely ignited the

(*Continued*)

BOX 2.1   THE SEMINAL CASES OF PHINEAS GAGE AND H.M. (*Continued*)

explosives with which he was working. The ensuing explosion sent the iron—which was 13 pounds in weight, 3 feet, 7 inches long in length, and 1 1/4 inch in diameter—through Gage's left cheek and out the top of his skull before landing approximately 80 feet behind him.

According to the account published by his treating physician, John M. Harlow, Gage "was thrown upon his back, and gave a few convulsive motions of the extremities, but spoke in a few minutes." His crew took the conscious Gage to a nearby hotel, where Harlow met him some hours later. Harlow reported that, upon arrival, he found Gage "to be perfectly conscious, but was getting exhausted from the hemorrhage, which was very profuse both externally and internally, the blood finding its way into the stomach, which rejected it as often as every 15 or 20 minutes."

Harlow went on to write that Gage made an "almost complete physical recovery." His report of Gage's physical status was also reflected in an evaluation conducted and reported by Dr. Henry J. Bigelow, then a prominent professor of surgery at Harvard Medical School. However, unlike Bigelow, Harlow reported sequela related to the injury to Gage's brain that caused significant changes to his personality and behavior:

> His contractors, who regarded him as the most efficient and capable foreman in their employ previous to his injury, considered the change in his mind so marked that they could not give him his place again. He is fitful, irreverent, indulging at times in the grossest profanity (which was not previously his custom), manifesting but little deference for his fellows, impatient of restraint or advice when it conflicts with his desires, at times pertinaciously obstinate, yet capricious and vacillating, devising many plans of future operation, which are no sooner arranged than they are abandoned in turn for others appearing more feasible. In this regard, his mind was radically changed, so decidedly that his friends and acquaintances said he was "no longer Gage."

Phineas Gage died in San Francisco in 1860, 12 years after his injury, from seizure-related complications. Harlow obtained consent from Gage's family to have his skull and the tamping iron preserved in the Warren Anatomic Museum at Harvard University School of Medicine, where it remains.

An autopsy was never performed on Gage. In the time since his death, psychologists, historians, and neuroscientists have attempted to

(Continued)

BOX 2.1   THE SEMINAL CASES OF PHINEAS GAGE AND H.M. (*Continued*)

retrace his life following the accident, and to use Gage's skull to reconstruct his injury and establish which areas of his brain were damaged. Although there are areas of disagreement, most have concluded that the significant damage occurred to the ventromedial prefrontal cortex (VMPFC), associated with social decision-making and self-control. Some have posited that this seminal case might also offer evidence of neuroplasticity, the brain's ability to recover lost functionality under certain conditions, as Gage appears to have been able to compensate for his deficits in later years through engagement in a highly structured career as a stagecoach driver in Chile.

Despite the lack of scientific certainty, the case of Phineas Gage firmly established that the brain and personality are linked and that brain damage, particularly to the prefrontal cortex (PFC), can influence behavior.

H.M.

In August 1953, patient H.M.—later revealed to be Henry Gustav Molaison—underwent bilateral medial temporal lobe resection surgery to resolve severe epileptic seizures. The then 27 year old may have acquired epilepsy following a head injury he sustained at age seven. By age 16, the seizures had become progressively worse. By 27, he was unable to work.

Undergoing the experimental surgery made sense.

The surgery was performed by Dr. William Beecher Scoville, a neurosurgeon at Hartford Hospital in Connecticut. Dr. Scoville removed the portion of the brain thought to be associated with the epilepsy, including parts of the hippocampus and amygdala.

At the conclusion of the surgery, H.M. could remember his name. He could remember his family members and his family history. He could remember the stock market crash of 1929, which he would have learned about as a child. The surgery significantly reduced his seizures.

H.M.'s memory for events leading to the surgery, however, was compromised. And, most significantly, HM suffered anterograde amnesia; he was unable to form new memories and, therefore, could not learn new words or faces. He could not remember his present life.

Prior to H.M.'s surgery, the full relationship between the hippocampus and memory formation was not known. After it, the bilateral removal of the hippocampi would not recur.

(*Continued*)

BOX 2.1 THE SEMINAL CASES OF PHINEAS GAGE AND H.M. (*Continued*)

Scoville consulted with eminent neurosurgeon Wilder Penfield and neuropsychologist Dr. Brenda Milner of the Montreal Neurological Institute, each of who were engaged in various memory experiments.

For the next 55 years, H.M. participated in select experiments that, for the first time, provided a more nuanced understanding of the way memories are formed and retained, as well as an enhanced understanding of the distinction between memory and other cognitive and perceptual abilities.

Despite his anterograde amnesia, for example, H.M. was able to sustain attention long enough to retain information for at least a brief period of time following its presentation. He could engage in uninterrupted conversations and continuously repeat a series of numbers for as long as 15 minutes. However, if H.M.'s attention was diverted, his memory for the conversation or task in which he had been engaged was lost. He was also unable to retain complex stimuli that were difficult to rehearse, such as faces. These findings led researchers to conclude that there is a fundamental distinction between short-term, or working memory, and long-term memory. Most significantly, the relevant factor between the two was not found to be time (i.e., the length of time H.M. was exposed to the stimuli did not correlate with whether he retained it in memory), but rather whether the information could be rehearsed.

An additional, significant finding was that H.M. was able to acquire "motor" or procedural memories, even in the absence of having a conscious or verbal memory of learning a particular task. In one experiment, Dr. Milner had H.M. repeatedly trace a five-pointed star under conditions where he could only observe his hand and the star in a mirror. H.M. retained the ability to draw the star, but could not articulate the process by which he learned to do so. This suggested a difference in the way in which procedural, or implicit, versus declarative, or explicit, memories are formed.

Over time, this led to the recognition that, while declarative memory is associated with the medial temporal lobe, other types of memories—such as those involved in skill, emotional, and perceptual learning—implicate regions including the amygdala, the basal ganglia, and the cerebellum.

H.M. died on December 2, 2008 at the age of 82, when his name was revealed.

His contribution to neuroscience generally, and to neurocriminology in particular, continues to reverberate, particularly in current research related to lie detection and victim memory (reviewed in Chapter 6).

The "Neuro" of Neurocriminology

The nervous system consists of the central nervous system (CNS) and the peripheral nervous system (PNS). The CNS consists of the brain and the spinal cord. The PNS includes the nerve fibers that connect the CNS to the body. The PNS is comprised of the somatic nervous system, which is responsible for transmitting sensory and motor information to and from the CNS, and the autonomic nervous system (ANS), which is responsible for involuntary bodily functions, such as breathing, digestion, heartbeat, and blood flow. The ANS is further divided into the sympathetic nervous system, which regulates "fight or flight" responses, and the parasympathetic nervous system, which conserves physical resources to support normal resting states.

The CNS is the mediator between our inner and outer worlds, regulating our processing of external stimuli and our responses to them. At the cellular level, the CNS is composed of approximately 100 billion neurons, and approximately ten times that number of supporting cells, which primarily consist of neuroglial cells, referred to as glial cells, or glial.

Each neuron contains a nucleus with the genetic material necessary for cell development and survival, and mitochondria that produce the energy that fuels cellular activity. Neurons also have specialized components called axons and dendrites, by which the brain communicates. Axons are responsible for information transmission. They are protrusions that arise from the soma (cell body) at the axon hillock and are generally covered with myelin, which facilitates rapid nerve impulse conveyance. Dendrites receive and process incoming information. Neurons do not touch; messages are communicated when an all-or-nothing electrical current, called an action potential, travels down the axon of a neuron and sends electrochemical messages— called neurotransmitters—across a synaptic gap, where the end of the pre-synaptic axon forms a terminal button to reach the receptor site of another neuron. These neuroreceptors are protein molecules embedded in the plasma of the neuron that selectively absorb specific neurotransmitters through one of two processes: a rapid post-synaptic change in the membrane or a slower, indirect electrical response.

The cell bodies, axon terminals, and dendrites of neurons comprise the brain's gray matter, which is pinkish-gray in color in the living brain. The axons, which connect the different gray matter regions, are the brain's white matter.

Neurons are divided into three classes depending upon their role: sensory neurons, motor neurons, and interneurons. Sensory neurons are responsible for mediating information in and outside the body, such as walking outdoors on a frigid day and knowing that the temperature is cold. Motor neurons obtain information and activate a motor response, such as to put on warmer

Brain Basics

clothing. Interneurons are involved in conveying information to and from other neurons, allowing for multiple responses, such as simultaneously experiencing the sensation of cold and donning warmer clothing. Interneurons are also implicated in learning and association, so that one would, for example, learn to anticipate the need to dress warmly when venturing outdoors in cold weather.

A single neuron can receive messages from many neurotransmitters. Some may be excitatory in nature, which call for firing an action potential, whereas others are inhibitory, which signal the action potential not to fire.

For a chemical to be classified as a neurotransmitter it must meet five criteria: (1) There must be evidence that it is synthesized in the presynaptic neuron. (2) The substance released from the nerve terminal should be in a chemically or pharmacologically identifiable form. (3) The specific changes in membrane properties seen after stimulation of the presynaptic neuron should be replicated at the postsynaptic cell. (4) The effects of neurotransmitter should be able to be blocked by competitive antagonists. (5) There should be active mechanisms to terminate the action of the neurotransmitter, including through pharmacological or enzymatic inactivation of the chemical messenger.

In 1914, acetylcholine (ACh) became the first identified neurotransmitter. Neurons that synthesize and release ACh are termed cholinergic neurons. Cholinergic neurons play a role in memory formation, verbal and logical reasoning, and concentration, and have been implicated in protection against neurodegenerative disease.

Since the identification of ACh, more than 100 additional neurotransmitters have been discovered.

The most common neurotransmitter in the CNS is glutamate, which is present in more than 80% of the brain's synapses, and is the main excitatory neurotransmitter. Neurons that respond to glutamate are referred to as glutamatergic neurons. Glutamatergic neurons are responsible for encoding information, forming and retrieving memories, spatial recognition, and maintaining consciousness. Excessive excitation of glutamate receptors has been associated with hypoxic injury, hypoglycemia, stroke, and epilepsy.

Gamma-aminobutyric acid (GABA) is present in the majority of other synapses, and in the retina. In the CNS, GABA is a major inhibitor of presynaptic transmission. Neurons that secrete GABA are termed GABAergic. GABAergic neurons are implicated in motor control, vision, and the regulation of anxiety.

Other neurotransmitters are present in fewer synapses, but have been found to be of significance in research related to criminal behavior. These include dopamine, norepinephrine (also referred to as noradrenaline), and serotonin.

Dopamine and norepinephrine, known as catecholamines, are involved in respiratory stimulation and psychomotor activity in the CNS, and have excitatory and inhibitory effects in the PNS.

The precursor to dopamine is the amino acid tyrosine, which is typically obtained through the diet. Tyrosine is chemically transformed into L-DOPA (L-3,4-dihydroxyphenylalanine), before becoming dopamine. Dopaminergic neurons are involved in mood, movement, planning, problem-solving, and learning.

Norepinephrine is synthesized from dopamine, and norepinephrine-secreting neurons are termed noradrenergic. These neurons are involved in controlling alertness, wake–sleep cycles, and in attention and memory.

Serotonin is a neurotransmitter predominantly found in the gastro-intestinal tract (90%), with the majority of the balance found in the CNS. Serotonin plays a significant role in the cardiovascular and respiratory system. In the CNS, serotonergic neurons play a key role in the regulation of anger and aggression, and of body temperature, mood, sleep, sexuality, and appetite. Serotonin is also involved in stimulation of the vomiting reflex.

The glial cells—named from the Greek word for "glue"—provide protection and support for the neurons. Four distinct types of glial cells are located between and around the neurons of the brain: astrocytes, oligodendrocytes, ependymal cells, and microglia.

> Astrocytes are responsible for transporting glucose and other substances out of the blood stream, and for processing glucose into lactate, a main energy source for neurons. Astrocytes also play a role in the uptake of certain neurotransmitters and in regulating extracellular potassium ion concentration, which are essential functions for the propagation of the action potentials by which neurotransmitters communicate information in the brain. Astrocytes are also involved in the process of forming neural scars after injury.

Oligodendrocytes form myelin, the membrane layer that covers the axon through which neurons transmit information, facilitating rapid communication.

Ependymal cells are found in the spinal cord and, although their function remains speculative, they are believed to be responsible for facilitating the transport of nutrients to the brain and the elimination of toxic metabolites from the brain.

Microglia are thought to play a key role during the development of the CNS, as well as to play a neuroprotective role in the event of injury or illness through removal of cellular waste and support of inflammation reduction.

Psychotropic and other medications, as well as illicit drugs, can impact brain functioning by increasing a neurotransmitter's excitatory or inhibitory effect (agonist) or by decreasing a neurotransmitter's excitatory or inhibitory effect (antagonist).

Brain Basics

Differences in how individuals think, perceive, remember, and act can be attributed to variations in the net total of transmissions across groups of neurons. Although brain functions are rarely isolated or confined to a specific region, neuroscience has begun to identify networks of neurons that correlate specific brain functions with specific areas of the brain. This, in turn, has informed efforts to identify brain functions associated with criminogenic behavior.

BOX 2.2   SUMMARY OF NEUROTRANSMITTERS IMPLICATED IN ANTISOCIAL AND CRIMINAL BEHAVIOR AND THEIR ROLE IN BRAIN FUNCTIONING

Neurotransmitter	Neuroreceptor	Role in Brain Functioning
Acetycholine	Cholinergic	Involved in autonomic ganglia, many autonomically innervated organs, at the neuromuscular junction, and at many synapses in the central nervous system. Involved in learning and memory.
Dopamine	Dopaminergic	Involved in nerve impulses in the substantia nigra and vental tegmental area. Involved in the control of locomotion, learning, working memory, cognition, and emotion. Abnormal functioning associated with various neurological and psychiatric disturbances such as Parkinson's disease, schizophrenia, and amphetamine and cocaine addiction.
Gamma-aminobutyric acid (GABA)	GABAergic	Major inhibitory neurotransmitter in the central nervous system. Implicated in anti-anxiety and relaxation.
Glutamate	Glutaminergic	Involved in neural communication, and in brain development, learning, memory, and cognitive function.
Norepinephrine/ noradrenaline	Adrenergic	Involved in sympathetic nervous system, increases in arousal and alertness, vigilance, formation and retrieval of memory, and focused attention.
Serotonin	Serotonergic	Involved in the regulation of mood, sleep, and appetite. Also involved in cognitive functions, memory, and learning.

The Structure and Functions of the Brain

The human brain can be rudimentarily divided into three main regions: the hindbrain, the midbrain, and forebrain.

The hindbrain, or rhombencephalon, is comprised of the medulla oblongata, the pons, and the cerebellum, and is responsible for basic life functions such as respiration, balance, coordination, movement of the limbs, and sleep–wake cycles.

The midbrain, or mesencephalon, is located in the brainstem and is composed of the tectum and tegmentum. It is responsible for visual and audio processing.

The forebrain, or prosencephalon, is divided into two parts: the diencephalon and the telencephalon.

The diencephalon includes the thalamus and hypothalamus. The thalamus serves as the main relay center between the basic life functions regulated by the medulla oblongata of the hindbrain, and the cerebrum, which is located in the telecephalon portion of the forebrain. The hypothalamus is associated with regulation of blood pressure and body temperature, thirst and hunger, pain and pleasure, and sex drive.

In addition to the cerebrum, the telencephalon includes basil ganglia. Basal ganglia are involved in refining movements to reach a particular goal, such swinging a baseball bat at a pitch. Basal ganglia are also associated with executing habitual actions, such as driving an automobile, and with learning new actions in novel situations.

The cerebrum is the largest, uppermost, and most highly developed part of the brain. It is the area responsible for cognition, perception, planning, and learning. It is divided into two hemispheres, the left and right. The hemispheres are connected by the corpus callosum, the largest fiber bundle in the brain, which is responsible for facilitating communication between the two hemispheres. The corpus callosum is also responsible for facilitating vision by combining images received by each hemisphere and connecting the visual cortex to the areas of the brain responsible for language to assist in object identification. The left hemisphere controls functions on the right side of the body, and the right hemisphere controls functions on the left side of the body. Generally, the left hemisphere of the brain controls speech and language functions, and is associated with processing information sequentially and logically. The right hemisphere controls visual–spatial functions, and is associated with processing information intuitively and holistically.

The Cerebral Cortex

The outer portion of the cerebrum—the cerebral cortex—is primarily comprised of a thin layer of gray matter (the neocortex) of approximately 1.5–5.5

Brain Basics

39

millimeters that, in humans, has six-folded layers that form the gyri (ridges) and sulci (groves). The layers contain distributions of neurons that form corical and subcortical connections associationally with other areas of the same hemisphere, or commissurally with the opposite hemisphere, primarily through the corpus callosum (see Box 2.3).

> **BOX 2.3 THE LAYERS OF THE CEREBRAL CORTEX**
>
> The neocortex is comprised of six-folded layers of gray matter that contain between 10 and 14 billion neurons. The layers are numbered using Roman numerals, with I representing the most superficial and VI representing the deepest.
>
> - Layers I–III represent the supragranular layers and are primarily responsible for intracortical connections.
>   Layer I is the molecular layer. It contains very few neurons and predominantly glial cells.
>   Layer II is the external granular layer. It contains small pyramidal (multipolar) neurons.
>   Layer III is the external pyramidal layer. It contains predominantly small and medium-sized pyramidal neurons and non-pyramidal neurons.
> - Layer IV is the internal granular layer and contains stellate (star-shaped) and pyramidal neurons. It receives thalamo-cortical connections, most prominent in the primary sensory cortices.
> - Layers V and VI are the infragranualar layers.
>   Layer V is the internal pyramidal layer and contains large pyramidal neurons which run to subcortical structures including the basal ganglia, brain stem, and spinal cord.
>   Layer VI is the multiform or fusiform layer and primarily projects into the thalamus.

The cerebral cortex is divided into four lobes: the occipital lobe, the temporal lobe, the parietal lobe, and the frontal lobe. Although neuroscience has demonstrated that there is often interdependence, redundancy, and overlap of functionality in the various regions of the brain, there are functions that are typically associated with each of the lobes (Photo 2.1).

Occipital Lobes

The occipital lobes are located at the very rear of the cerebral cortex, and house the visual cortex. The occipital lobes are responsible for visual

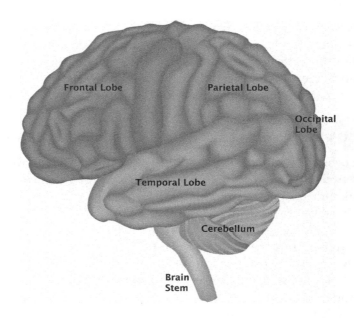

reception, visual-spatial processing and interpretation, and color and motion recognition—that is, in directing visual awareness to important features in the environment.

Damage to the occipital lobes can lead to visual impairments, including difficulty recognizing or naming objects, visual hallucinations, or blindness.

Temporal Lobes

The temporal lobes are situated just above the ears on the lower-middle sides of the cerebral cortex. The temporal lobes contain the primary auditory cortex and are responsible for processing all auditory information. The posterior region of the left temporal lobe contains Wernicke's area, which is responsible for receptive language. The temporal lobes also house the hippocampus, which is responsible for the consolidation of new, long-term verbal and visual memories.

Damage to the temporal lobes can lead to memory problems, as well as problems with speech perception, hearing, and receptive language. Left temporal lobe damage can result in challenges in audio and visual recall, word recognition, and memory for verbal material. Right temporal lobe damage can result in problems recognizing or remembering visual or audio content.

Parietal Lobes

Located behind the frontal lobes, in the middle section near of the crown of the skull, the parietal lobes includes the somatosensory cortex, which is responsible for receiving and interpreting sensory information from various parts of the body. The parietal lobes are also responsible for attention, spatial

Brain Basics

orientation, reading, and voluntary motion, as well as complex visual processing, such as mentally rotating three-dimensional shapes.

Damage to the left parietal lobe can result in aphasia (language disorder), agnosia (abnormal object perception), or Gertsmann's syndrome, which is characterized agraphia or dysgraphia (writing disability), acalculia or dyscalculia (a lack of understanding of the rules for calculation or arithmetic), and finger agnosia (an inability to distinguish right from left, and an inability to identify fingers). Damage to the right parietal lobe can also result in impaired personal care and impaired drawing ability. Bilateral parietal lobe damage can result in Balint's syndrome, which is characterized by an inability to perceive the visual field as a whole (simultanagnosia), difficulty in fixating the eyes (oculomotor apraxia), and inability to move the hand to a specific object by using vision (optic ataxia).

Frontal Lobes

The frontal lobes are located in the front portion of the cerebrum, right behind the forehead. The frontal lobes are inferiorly separated from the temporal lobes by the Sylvian fissure and posteriorly separated from the parietal lobes by the central sulcus. They are the largest in the human brain, are the most recently developed, and are the most highly evolved. The frontal lobes are responsible for initiation of all voluntary movement, including speech production. The lower left frontal lobe contains Broca's area, which is responsible for expressive language.

The anterior portions of the frontal lobes are called the PFC. The PFC is responsible for higher cognitive or executive functions—what have been referred to as "top–down processing"—or the application of cognition to perception or stimuli to inform response.

The prefrontal lobes can be referenced topographically, that is, dorsal/superior (upper); ventral/inferior (lower); rostral/interior (front); caudal/posterior (back) and directionally, that is, medial (front to back) and lateral (left to right).

The dorsolateral prefrontal cortex (DLPFC) has extensive neural connections to sensory and motor cortices and is involved in the regulation of action, thought, and attention.

The ventromedial prefrontal cortex (VMPFC) has extensive connections to the limbic system, discussed below, and is engaged in the regulation of emotional responses.

Additionally, the PFC includes the orbitofrontal cortex, which is involved in goal selection and the ability to understand and evaluate future rewards.

The PFC regions collaborate to support recognition of similarities and differences between things and events, evaluation and inhibition of socially unacceptable behavior, and choice between good and bad options— independent of external or internal distractions or stimuli. These capacities, in turn, support reality-testing, sound judgment, and the ability to course correct if errors are made, each of which has clear ramifications for potential engagement in criminal behavior.

The Limbic System

Below the cortical surface are more primitive brain structures that are essential for survival and which are collectively called the limbic system. The limbic system is called "the emotional brain"; it is responsible for emotional responsiveness, formation and consolidation of memories, olfaction, and motivation. The limbic system is associated with what has been referred to as "bottom–up processing" or the experience of information or stimuli based on sensory perception.

Although there is some debate as to what comprises the subcortical limbic brain, this system is generally thought to include the hypothalamus, the hippocampus, the amygdala, and the cingulate cortex (Photo 2.2).

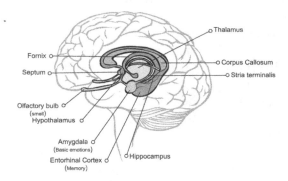

Hypothalamus

The hypothalamus is responsible for certain metabolic functions of the ANS, such as those that control body temperature, hunger, thirst, and sleep. The hypothalamus also plays a role in emotion, particularly in relation to aversion and displeasure. It appears to play a critical role in triggering fear responses in relation to external stimuli. The hypothalamus has also been implicated in aspects of parenting and attachment behaviors. Lesions of the hypothalamus have been associated with changes in sexuality, combativeness, and hunger.

Hippocampus

The hippocampus is generally associated with the memory formation. Damage to the hippocampus usually results in anterograde amnesia, or difficulty in forming new memories, and can also result in retrograde amnesia, or difficulty in accessing memories formed prior to the insult.

Brain Basics

Amygdala

The amygdala plays a key role in emotional processing and responding, and informs the body's detection of threatening stimuli or engagement in a "fight-or-flight" reaction. Variations in amygdala functioning have been implicated in many psychological disorders, including obsessive compulsive disorder, posttraumatic stress disorder, borderline personality disorder (PD), and bipolar disorder, as well as in the dysregulation of aggressive behaviors, including hyperaggressiveness. Damage or removal of the amygdala has been associated with Klüver-Bucy syndrome, which is characterized by docility, even in the face of threat, as well as hypersexuality and hyperorality.

Cingulate Cortex

The cingulate cortex is located above the corpus callosum and plays an important role in linking sensation, emotion, and action, including in relation to the formation of long-term memories for emotionally-significant events and dealing with uncertainty. Dysfunction related to the cingulate cortex has been implicated in apathy, depression, and schizophrenia.

BOX 2.4 SUMMARY OF BRAIN REGIONS AND GENERALLY ASSOCIATED FUNCTIONS

Frontal lobe	Responsible for initiation of all voluntary movement, including speech production. The frontal lobes include the prefrontal cortex, which is responsible for higher cognitive or executive functions.
Parietal lobe	Responsible for receiving and interpreting sensory information from various parts of the body. Also responsible for attention, spatial orientation, reading, and voluntary motion, as well as complex visual processing, such as mentally rotating three-dimensional shapes.
Temporal lobe	All auditory information. Also house the hippocampus, which is responsible for the consolidation of new, long-term verbal and visual memories.
Occipital lobe	Visual reception, visual–spatial processing and interpretation, and color and motion recognition (directing visual awareness to important features in the environment).
Limbic system	The "emotional brain." Generally thought to include the hypothalamus (responsible for controlling autonomic functions including body temperature, hunger, thirst, and sleep), the hippocampus (involved in memory), the amygdala (which plays a key role in emotional processing and responding), and the cingulate cortex (which plays an important role in linking sensation, emotion, and action, including in relation to the formation of long-term memories for emotionally-significant events).

Brodmann's Areas

One of the common systems used to classify the neuronal systems of the cerebral cortex was developed in 1909 by German anatomist Korbinian Brodmann. Based upon his cell staining studies involving epileptic patients, monkeys and other primates, Brodmann published cortical maps that remain the most commonly cited cytroarchitectural system. The areas correlate with the psychological and behavioral functionality Brodmann identified in each of the lobes of the cerebral cortex (Table 2.1):

Table 2.1 Brodmann's Areas

Frontal Lobe	BA4	Precentral gyrus/primary motor area
	BA6	Premotor area and the supplementary Motor area.
	BA8	Facilitates eye movements. Involved in inductive reasoning and planning, as well as memory processes, particularly working memory and sequence learning.
	BA9, 10	Involved in memory encoding and retrieval and working memory. BA10 may be involved in controlling memory ("metamemory" including intentional forgetting).
	BA11	Associated with reaction style or "personality."
	BA44, 45	Broca's area. Involved in expressive language, verbal fluency and sequencing, verbal working memory, selecting information among competing sources.
Parietal Lobe	BA3, 2, and 1	Primary somatosensory area, associated with touch and proprioception, including kinesthesia, and mirror neurons, which are active when observing others and may be implicated in anticipating behavior and social learning.
	BA5, 7	Presensory association areas. Implicated in somatosensory processing, including perception of personal space.
	BA40	Involved in complex verbal processing, including verbal creativity, as well as deductive reasoning and engagement in creative tasks.
	BA39	Angular gyrus. Involved in associations between somatosensory information, auditory information, and visual information. Implicated in calculation, reading/writing, and naming abilities. Also involved in verbal creativity, inferential reasoning, and sequencing.
Temporal Lobe	BA41	Heschl's gyrus (primary auditory area).
	BA42	Involved in more detailed detection and recognition of speech.
	BA21, 22	Wernicke's area. Involved in receptive language processing.
	BA37	Involved in associating words with visual stimuli.
Occipital Lobe	BA17	Primary visual area.
	BA18, 19	Secondary visual areas; where visual associations and processing occur.

Brain Basics 45

Common Causes of Brain Dysfunction

The brain is protected from injury and illness by the cranium, meninges, cerebrospinal fluid, and the blood–brain barrier. The cranium, or skull, includes the bones that enclose the brain. The meninges are beneath the surface of the skull and are comprised of three membrane layers—the dura, arachnoid, and pia mater—that surround the brain and spinal cord. Cerebrospinal fluid is the watery, clear substance that is nearly devoid of cells and is located within the subarachnoid space in the central canal of the spinal cord and the four ventricles of the brain. The fluid is formed continuously in the ventricles to maintain constant pressure of approximately 100–180 millimeters of water in a side-lying adult. The cerebrospinal fluid helps protect the brain, spinal cord, and meninges by absorbing any shocks to which they are exposed. The blood–brain barrier is a selectively permeable structural and functional cell wall, comprised in part of astrocytes, that allows entry of substances such as glucose, ions, and oxygen, while preventing entry of most blood-borne toxins.

Despite these physiological protections, the brain is susceptible to diverse factors that can contribute to dysfunction. Common causes of cerebral dysfunction include tumors, stroke, traumatic brain injury (TBI), epilepsy, neurodegenerative disease, psychotic disorders, PDs, and substance abuse.

Tumor

According to the World Cancer Research Fund International, the incidence rate of malignant and nonmalignant brain tumors worldwide from the latest available reporting period (2012) is 3.4 per 100,000, with developed countries reporting rates of 5.1 per 100,000 and less developed countries reporting rates of 3.0 per 100,000.

Brain tumors are generally characterized based upon the location of the tumor, the type of tissue involved, and whether the tumor is malignant (cancerous) or benign (noncancerous). Tumors are often graded along a scale of one to four, based upon level of malignancy (see Box 2.2). A brain tumor is referred to as primary if it originates in the brain cells, or metastatic or secondary if it results from cancer cells that originated in another part of the body and have spread to the brain.

Brain tumors can damage the brain by directly destroying brain cells, by producing inflammation that places pressure on other parts of the brain, or by increasing pressure within the skull.

In adults, gliomas tumors are the most common:

- Astrocytomas. Can be any grade, and are often associated with seizures or changes in behavior.

46   Neurocriminology

- Meningiomas. The most common primary brain tumors in adults. Most likely to occur in adults age 70–80. Arise in the meninges. Can be grade 1, 2, or 3. Are usually benign and grow slowly.
- Oligodendrogliomas. Are usually grade 1, 2, or 3. Typically grow slowly and don't spread to nearby tissue.

Brain and spinal cord tumors are the second most common cancers in children, accounting for approximately one in four childhood cancers. Brain tumors in children are more likely to start in the lower parts of the brain, such as the cerebellum and brain stem.

Brain tumors, particularly those that implicate the prefrontal cortex or limbic system, have been associated with emotional dysregulation and increased aggression.

BOX 2.5   BRAIN TUMOR MALIGNANCY SCALE

Brain tumor malignancy is generally rated on a scale of 1–4:

- Grade 1 cells look nearly normal and grow slowly, and long-term survival is considered likely.
- Grade 2 cells look slightly abnormal and grow slowly. The tumor may spread to nearby tissue and can recur later, maybe at a more life-threatening grade.
- Grade 3 cells look abnormal and are typically growing into nearby brain tissue. The tumors tend to recur.
- Grade 4 cells look abnormal and grow and spread quickly.

Stroke

The World Health Organization estimates for stroke incidences vary from 50 to 500 per 100,000 depending on region. The vast majority of strokes are ischemic strokes, which result from a clot in the artery that supplies the brain with oxygen-rich blood. A hemorrhagic stroke occurs when an artery in the brain leaks blood or ruptures, placing pressure on brain cells and resulting in brain damage. High blood pressure and aneurysms are common causes of hemorrhagic strokes. Hemorrhagic strokes can be intracerebral, resulting when an artery within the brain bursts and the surrounding tissue is flooded with blood, or subarachnoid, which occurs when the bleeding occurs between the brain and the arachnoid mater.

As with brain tumors, strokes have been implicated in hyperaggressiveness, emotional dysregulation, and increased impulsivity in some individuals, depending upon the area of the brain effected.

Brain Basics

Trauma

TBI represents a continuum of acquired or sudden brain insults that can be characterized as mild, moderate, or severe and accompanied by no, limited, or prolonged loss of consciousness. Types of TBI include:

- Skull fracture: Occurs when the skull cracks. Pieces of broken skull may cut into the brain and injure it.
- Contusion: Occurs when the brain is bruised. Swollen brain tissue mixes with blood released from broken blood vessels.
- Intracranial hematoma: Occurs when damage to a major blood vessel in the brain or between the brain and the skull causes bleeding.
- Anoxia: Occurs when there is an absence of oxygen to the brain that causes damage to the brain tissue.

TBI is most common in children under 4 years old, in young adults between 15 and 25 years old, and in adults 65 and older. Half of all TBIs result from motor vehicle accidents. Other causes include bullet wounds, shaken baby syndrome, sports-related injuries, and military incidents.

Although the majority of individuals who suffer TBIs do not engage in criminal behavior, TBIs suffered in childhood have been associated with the development of antisocial PD. Retrospective research on violence criminals has also found a correlation with a history of TBI, particularly to the prefrontal cortex (see Box 2.3).

BOX 2.6   CTE AND THE CASE OF AARON HERNANDEZ

TBI refers to a continuum of conditions that can be mild, moderate, or severe, and have a myriad of causes.

One type of TBI is chronic traumatic encephalopathy, or CTE, a condition associated with athletes, military personnel, and others subject to repetitive brain trauma.

The clinical presentation of CTE was first described in 1928 by Dr. Harrison Martland to characterize boxers who exhibited what he referred to as "punch drunk syndrome."

More than three-quarters of a century later, neuropathologist Bennet Omalu and his colleagues at the University of Pittsburg published CTE-related findings on former Pittsburgh Steeler football player Mike Webster. Webster died suddenly at the age of 50, after years of struggling with cognitive and intellectual impairment, mood disorders, drug abuse, and suicide attempts. Omalu's analysis of Webster's brain tissue at autopsy found large accumulations of tau protein in Webster's

(Continued)

BOX 2.6   CTE AND THE CASE OF AARON HERNANDEZ (*Continued*)

brain, which is associated with mood dysregulation and compromise to executive functioning.

Three years after the publication of Omalu's findings, the VA-BU-CLF Brain Bank—a partnership between the Concussion Legacy Foundation, Boston University, and the Veteran's Administration—formed to expand knowledge of CTE and brain trauma.

In 2017, the concern examined the brain of another deceased football player, former New England Patriots Aaron Hernandez, and declared that suffered Stage 3 CTE.

The most severe form of the disease, which currently can only be diagnosed during autopsy, is characterized as Stage 4.

Evidence suggests that CTE is caused by a significant number of repetitive concussive or, more typically, sub-concussive impacts to the head sustained over a period of time. Initial symptoms of CTE, which do not generally appear until years following the onset of head impacts, affect mood and behavior, and include depression, impulsivity, aggression, and paranoia. Cognitive symptoms, including memory loss, impaired judgment, and progressive dementia, typically present in the later stages of the disease.

Aaron Hernandez was a high-school football star who reportedly became involved with drugs and violence following the unexpected death of his father during a routine surgery. Despite this, he went on to play for the University of Florida and was a fourth round Patriot draft pick.

Hernandez's success at the Patriots resulted in his signing a 7-year, $40 million contract after the 2012 season. In the summer of 2013, Hernandez was arrested and charged with the murder of Odin Lloyd, a semiprofessional player who was dating the sister of Hernandez's fiancée. Lloyd's body was found a mile away from Hernandez's home.

A jury convicted Hernandez of the killing in 2015. During the trial, he was indicted for the 2012 murders of Daniel de Abreu and Safiro Furtado, two strangers to Hernandez who the state argued that he killed outside a Boston club.

The 27-year-old Hernandez hanged himself in his cell 4 days after a jury had acquitted him of the 2012 murders, and while serving a sentence of life without the possibility of parole for the 2015 murder.

At a conference held months following her examination of Hernandez's brain and after receiving the permission of Hernandez's

(Continued)

Brain Basics 49

> ### BOX 2.6   CTE AND THE CASE OF AARON HERNANDEZ (*Continued*)
>
> family, Dr. Ann McKee, neuropathologist and director of the CTE center, reported that Tau protein was found in Hernandez's frontal cortex, the amygdala, and the temporal lobe. She stated that the median age of a Stage 3 brain from individuals involved in football was 67. McKee added that Hernandez had a genetic marker that may have rendered him vulnerable to certain brain diseases, and could have contributed to aggressive progression of the disease.
>
> She cautioned the following:
>
> We can't take the pathology and explain the behavior. But we can say collectively, in our collective experience, that individuals with CTE—and CTE of this severity—have difficulty with impulse control, decision-making, inhibition of impulses for aggression, emotional volatility, rage behaviors. We know that collectively.
>
> Her words perfectly reflect the contemporary dialogue related to the translation of neuroscience to the criminal justice system, explored in greater detail in Chapters to follow.

Epilepsy

According to the Epilepsy Foundation, epilepsy is the fourth most common neurological condition, and affects an estimated 65 million individuals worldwide. Epilepsy is a chronic condition, characterized by disorganized electrical events (seizures) in the brain. It is diagnosed when an individual exhibits two unprovoked seizures, or one seizure with a high probability of another occurring, that are not caused by a known and reversible condition such as alcohol withdrawal or extremely low blood sugar.

When seizures appear to involve groups of cells on both sides of the brain they are referred to as generalized seizures. Generalized seizures include:

- Absence seizures: Previously known as petit mal seizures. Often occur in children. Are characterized by staring into space or by subtle body movements such as eye blinking or lip smacking. May occur in clusters. May cause a brief loss of awareness.
- Tonic seizures: Cause muscle stiffening, usually in the back, arms and legs. May result in falling.
- Atonic seizures: Also known as drop seizures. Cause loss of muscle control. May result in sudden collapse.

50 Neurocriminology

- Clonic seizures: Associated with repeated and rhythmic muscle movements. Usually affect the neck, face, and arms.
- Myoclonic seizures: Usually sudden, brief twitches of arms and legs.
- Tonic–clonic seizures: Previously known as grand mal seizures. Can cause abrupt loss of consciousness, body stiffening and shaking, loss of bladder control and tongue biting.

Focal (partial) seizures result from abnormal brain activity in one area of the brain. Focal seizures are generally subtyped by whether they are accompanied by a loss of awareness. Focal seizures without a loss of consciousness (formerly referred to as "simple partial seizures") may nonetheless alter emotions and sensory perception, as well as result in involuntary movement. Focal seizures with impaired awareness (formerly referred to as "complex partial seizures") involve a loss of consciousness or significant alteration in awareness that can include non-responsiveness to environmental stimuli, or repetitive movements such as hand rubbing, chewing, or walking in circles.

Seizures generally occur in phases. The prodormal stage can occur days or hours prior to the seizure. It is typically characterized by changes in mood, affect, or cognition. The aura stage occurs seconds before the seizure onset. Depending upon the area of the brain affected, aura symptoms can include abnormal sensations including déjà vu (familiar feelings) or jamais vu (unfamiliar feelings), dizziness, headache, numbness, lightheadedness, nausea, hallucinations, and emotional dysregulation. The ictus stage is the seizure itself. The postictal stage is the period of recovery.

Most research on the association between epilepsy and aggression or violence suggests that it occurs during the ictus stage, or the actual seizure, and that it is neither intentional nor directed.

Neurocognitive Disorders

Major Neurocognitive Disorder due to Alzheimer's disease (formerly Dementia of the Alzheimer's type) results from decreases in the ACh production of neurons in the cerebral cortex and subcortical regions, which manifests as progressive memory loss, changes in personality and challenges in executive functioning. Violence and aggression have also been associated with Alzheimer's disease; these physical manifestations are thought to result from disorientation and misreading the environment, delusions or visual hallucinations, and/or misperceiving possible threats.

In contrast, frontotemporal dementia (FTD) is a term for a group of neurodegenerative diseases of the frontal and temporal lobes not caused by Alzheimer's disease. According to the National Institutes of Health, FTD is one of the leading causes of early onset dementia. Thirteen percent of FTD

Brain Basics

cases occur in individuals younger than 50 years of age, and the mean age of onset is 56. Prevalence of FTD is estimated at between 15 and 22 per 100,000.

FTD is significant for its symptomology; patients present with disinhibition and impulsivity prior to exhibiting cognitive or memory challenges. Consequently, their personality and behavioral changes place them at higher risk for misdiagnosis, preempting receipt of the supervised care needed to manage the condition. The behavioral and emotional dysregulation associated with the early stages of the disease has been associated with engagement in high-risk activities, including crime. A 2015 study conducted by Liljegren and colleagues involving more 2,000 patients at University of California, San Francisco Memory and Aging Center found that patients with behavioral variant FTD were significantly more likely to have engaged in criminal behavior following the likely time of onset of the disease compared with patients with Alzheimer's dementia (14% versus 2%). Additionally, these patients were 6.4% more likely to exhibit violence compared with 2% of patients with Alzheimer's dementia.

Psychotic and Mood Disorders

Psychotic disorders are characterized by a distorted perception of reality. The most common form of psychosis, schizophrenia, has an approximate prevalence rate of 1% worldwide among people 18 years or older. Schizophrenia may include positive symptoms such as delusions and hallucinations, and negative symptoms such as withdrawal, lack of motivation, and apathy.

Structural dysfunctions that have been associated with schizophrenia include nonlocalizcd reduced gray matter and white matter changes, temporal lobe volume reductions, and anomalies of the temporal and frontal lobe white matter. Abnormally high levels of glutamate and of dopamine have been generally implicated in schizophrenia. Additionally, decreased relative prefrontal cortex glucose metabolism (hypofrontality) has been associated with negative symptoms.

Studies on delusional disorder—in which individuals typically have the single psychotic symptom of a fixed abnormal belief—show some shared structural and functional variants with schizophrenia, and some significant differences. Those with delusional disorder, for example, had less widespread reductions in cortical gray matter. Additionally, those with delusional disorder did not exhibit the hypofrontality that was generally found in functional imaging conducted on individuals with schizophrenia.

There has been some evidence to suggest an association between schizophrenia and violence, particularly when substance abuse and acute positive symptomology, such as command hallucinations, are present. There also has been evidence of an association between delusional disorder and violence, particularly when paranoid and erotomania subtypes are included. However, the overall rates of violence, particularly severe violence such as homicide, by those with schizophrenia and delusional disorder are statistically low when compared with the general population.

Similar conclusions have been reached in relation to those with various mood disorders; although mood disorders have been implicated in certain crime subtypes—such as major depression in instances of homicide–suicide and filicide—most individuals who suffer from mood disorders do not engage in violence or commit crimes.

Meta-analysis on structural and functional brain imaging studies that have explored correlates between neural circuitry and mood disorders have consistently implicated the frontal lobe, limbic system, and basal ganglia circuits, and have found serotonin and dopamine to play critical roles.

Personality Disorders

PDs are enduring, maladaptive patterns of relating to oneself and others that affect an estimated 12% of the population. Those with PDs generally suffer from disordered cognition, distorted affect, impulsivity, and interpersonal difficulties. The ten PDs are clustered by shared characteristics. Cluster A PDs are characterized by odd or eccentric behaviors and include Schizoid Personality Disorder, Paranoid Personality Disorder, and Schizotypal Personality Disorder. Cluster B PDs are characterized by dramatic or irrational behavior and include Antisocial Personality Disorder, Borderline Personality Disorder, Narcissistic Personality Disorder, and Histrionic Personality Disorder. Cluster C's are characterized by anxious or fearful behaviors and include Dependent Personality Disorder, Obsessive-Compulsive Personality Disorder, and Avoidant Personality Disorder.

The majority of research related to the association between criminality and PDs focuses upon Cluster B disorders. Current research suggests that, as opposed to general clinical samples, Cluster B disorders, particularly antisocial and borderline PDs, is especially prevalent in forensic psychiatric samples, reaching nearly 80% in women, and is strongly associated with degree of violence severity in offenders of both genders.

Structural brain imaging conducted on individuals with borderline PD has found significant decreases in hippocampal and amygdala volume compared to control subjects, as well as significant volume reduction of the left orbitofrontal cortex and the right anterior cingulate cortex. Additionally, antisocial PD has been associated with lower prefrontal activity and higher subcortical activity.

In addition to these PDs, psychopathy, which has been recognized in research, clinical, and forensic arenas for decades, was added to the most recent addition of the American Psychiatric Association's *Diagnostic and Statistical Manual of Mental Disorders*, or DSM-5 (although it is included in Section III and is, therefore, not intended for general clinical use). Psychopaths

Brain Basics

demonstrate a shallow, callous, and manipulative interpersonal style that is combined with antisocial, reckless, and frequently violent behavior. Studies have found that, like those with PDs, psychopaths exhibit higher subcortical activity. Unlike those with PDs, however, psychopaths do not typically exhibit compromises to the prefrontal lobes, which is consistent with their engagement in more predatory or planned violence.

Substance Abuse

The U.S. Department of Justice estimates that more than half of those incarcerated in prisons and jails are dependent upon or abuse drugs—ten times the 5% prevalence rate for the general population.

The impact of drug and alcohol use and abuse on brain functioning, and on criminal behavior, is particularly complex. Illicit drug use can, of itself, be characterized as criminal depending upon the drug, or the context in which the drug is used (e.g., driving while under the influence of alcohol). Drug use can also motivate criminal activity, either by engagement in secondary crimes as a means to procure the drug of choice, or as a result of the impact on brain functioning and, subsequently, on behavior.

According to Substance Abuse and Mental Health Services Administration (SAMHSA), alcohol is the most abused substance. In addition, an estimated 24.6 million of Americans age 12 or older, representing over 9% of the population, reported using illicit drugs. The overwhelming majority reported using marijuana (19.8 million), considered an illicit drug under the federal law as of 2017, and/or prescription drugs (6.5 million), defined as illicit when misused. One and one-half million individuals (1.5 million) reported using cocaine in the prior month, making it the third most abused illicit drug.

Numerous studies have found an association between alcohol consumption and aggression. Alcohol increases the effects of GABA and dopamine, while inhibiting glutamate, which account for alcohol's sedating effects. Recent studies on blood alcohol content (BAC) levels have shown that alcohol also impacts norepinephrine. Specifically, rising BAC levels that occur during drinking are correlated with elevated levels of norepinephrine, which is associated with heightened arousal and impulsivity. Additionally studies have found that alcohol consumption is associated with decreased activity in the prefrontal cortex. The increased tendency toward impulsivity, coupled with compromises to the executive functioning and emotional and behavioral dsyregulation, may be contributory factors in the correlation between alcohol abuse and criminality.

There have been conflicting findings in imaging studies on brain structure and functioning and marijuana, the second most used drug in America and the most used in many parts of the world. Some studies suggest that consistent marijuana use in adolescence can result in cognitive impairments, including decreases in IQ of up to 8 points, as well as adverse impacts on learning and memory. Other studies have found no significant differences in brain structure between users and nonusers.

A 25-year longitudinal study that utilized data from more than 3,000 participants of the Coronary Artery Risk Development in Young Adults found that past exposure to marijuana is associated with worse verbal memory but does not appear to affect other domains of cognitive function, such as processing speed or executive functioning.

More recently, a 50-year study followed a cohort in London and controlled for the often confounding variables of family history of criminality, childhood antisocial behavior, mental health history, and alcohol and other drug use. The researchers concluded that there is a causal association between persistent cannabis use and violence.

Research has also suggested that marijuana may serve as a "gateway" drug, leading to use of other illicit substances. Studies in this area are predominantly based on animal research. Early exposure to cannabinoids in adolescent rodents, for example, was found to decrease the reactivity of dopamine in adulthood. Additionally, rats who were administered THC were found to be demonstrate receptivity to other drugs, known as cross-sensitization.

The legalization of marijuana in several states in the U.S. has raised questions and concerns regarding the potential impact on crime. Research to date has primarily focused upon states that have legalized marijuana for medicinal purposes; states that have legalized medical marijuana have seen proportionately lower crime rates than the national average since legalization. As research has not typically controlled for confounding variables, such user age or persistency of use, it is unclear as to whether these findings contradict previous studies, or will generalize to legalization for recreational use.

The long-term use of pain medication, particularly opiods, has been associated with dysfunctions of the limbic system and the orbitofrontal cortex, and associated compromises in executive functioning, cognition, and emotional regulation. However, the association between opiod use and engagement in violence or secondary crimes has been mixed; some studies cite no correlation, and others find no significant correlation beyond that associated with substances generally. One large-scale study that focused on over 10,000 adolescent participants of the Washington State Health Youth Survey employed an innovative design that included violent ideation as well as own-prescription and diverted opiod use to control for situational factors, such as engagement with antisocial peers or engagement in violence to procure the drug. The study found that, controlling for these factors, opiod abusing teens had a significantly higher likelihood of violence engagement, implying a relationship between opiod use and violence that transcends the environmental factors previously posited.

Brain Basics

Cocaine and other stimulants, including methamphetamine, affect dopamine, serotonin and norepinephrine, and therefore have a particularly significant impact on mood, cognitive function, and impulsivity. Specifically, cocaine blocks receptors responsible for removing dopamine from synapses, effectively exaggerating and prolonging pleasurable neural signals. Conversely, methamphetamine increases the release of dopamine, as well as serotonin and norepinephrine. Each has been associated with hyperaggressiveness and violence.

Alcohol and drugs mimic neurotransmitters and have cumulative physiological effects that motivate continued drug use by co-opting the body's natural process for producing rewarding dopamine. This process, coupled with the aversive consequences of substance withdrawal, render addiction difficult to overcome and a significant contributor to brain dysfunction.

Through extensive research, modern science has illuminated much about the way the brain functions. With the advent of neuroimaging, science has the opportunity to expand our knowledge base of this incredibly complex organ, and to potentially contribute to deepening our understanding of correlates to criminal behavior.

Key Terms

Central Nervous System (CNS): Part of the nervous system comprised of the brain and the spinal cord that is responsible for processing and responding to stimuli.

Peripheral Nervous System (PNS): Part of the nervous system that includes the nerve fibers that connect the central nervous system to the body. The PNS is comprised of the somatic nervous system, which is responsible for transmitting sensory and motor information to and from the CNS, and the autonomic nervous system (ANS), which is responsible for involuntary bodily functions, such as breathing, digestion, heartbeat, and blood flow.

Neurons: Tissue responsible for transmitting and receiving nerve impulses. Includes axons that are responsible for information transmission and dendrites that receive and process incoming information.

Glial: The connective tissue of the nervous system that protects neurons and supports neurotransmission.

Myelin: The insulating sheath around many nerve fibers that support neurotransmission of information.

Neurotransmitters: A chemical substance across the synapse that causes the potential transfer of an impulse to another neuron.

Gray Matter: The tissue of the brain that consists primarily of the cell bodies, axon terminals, and dendrites.

White Matter: Axons, typically covered in mylenated sheaths, which connect the different gray matter regions.

Agonists: Agent that can impact brain functioning by increasing a neurotransmitter's excitatory or inhibitory effect.

Antagonist: Agent that can impact brain functioning by decreasing a neurotransmitter's excitatory or inhibitory effect.

Hindbrain: Also known as the rhombencephalon. Includes the medulla oblongata, the pons, and the cerebellum. Is responsible for basic life functions such as respiration, balance, coordination, movement of the limbs, and sleep–wake cycles.

Midbrain: Also known as the mesencephalon. Located in the brainstem and is composed of the tectum and tegmentum. Is responsible for visual and audio processing.

Forebrain: Also known as the prosencephalon. Divided into the diencephalon, which includes the thalamus and hypothalamus, and the telencephalon, which includes the cerebrum and basil ganglia.

Thalamus: Main relay center between the basic life functions regulated by the medulla oblongata of the hindbrain, and the cerebrum.

Hypothalamus: Responsible for regulation of blood pressure and body temperature, thirst and hunger, pain and pleasure, and sex drive.

Basal Ganglia: Group of structures involved in refining movements to reach a particular goal, with executing habitual actions, and with learning new actions in novel situations.

Cerebrum: Largest, uppermost, and most highly developed part of the brain. Responsible for cognition, perception, planning, and learning. Divided into left and right hemispheres.

Corpus Callosum: The fiber bundle in the brain responsible for facilitating communication between the right and left hemispheres of cerebrum, as well as for facilitating vision by combining images received by the visual cortex.

Occipital Lobes: Located at the very rear of the cerebral cortex. House the visual cortex, and are responsible for visual reception, visual–spatial processing and interpretation, and color and motion recognition.

Temporal Lobes: Located above the ears on the lower-middle sides of the cerebral cortex. House the primary auditory cortex, and are responsible for processing all auditory information. Also house the hippocampus, which

Brain Basics

is responsible for the consolidation of new, long-term verbal and visual memories.

Parietal Lobes: Located behind the frontal lobes near of the crown of the skull. House the somatosensory cortex, which is responsible for receiving and interpreting sensory information from various parts of the body. Also responsible for attention, spatial orientation, reading, and voluntary motion, as well as complex visual processing, such as mentally rotating three-dimensional shapes.

Frontal Lobes: Located in the front portion of the cerebrum, behind the forehead. Responsible for initiation of all voluntary movement, including speech production.

Limbic System: Located below the cortical surface. Called "the emotional brain." Responsible for emotional responsiveness, formation and consolidation of memories, olfaction, and motivation. Generally thought to include the hypothalamus, the hippocampus, the amygdala, and the cingulate cortex.

Hypothalamus: Responsible for certain metabolic functions of the autonomic nervous system, such as those that control body temperature, hunger, thirst, and sleep. Also plays a role in emotion and in triggering fear responses in relation to external stimuli.

Hippocampus: Responsible for memory formation.

Amygdala: Responsible for emotional processing and responding. Informs the body's detection of threatening stimuli or engagement in a "fight-or-flight" reaction.

Cingulate Cortex: Responsible for linking sensation, emotion, and action, including in relation to the formation of long-term memories for emotionally significant events.

Use Your Brain

Test Your Knowledge

1. The autonomic nervous system (ANS) is divided into two parts:
    a. The sympathetic nervous system, which regulates "fight or flight" responses, and the parasympathetic nervous system, which conserves physical resources to support normal resting states.
    b. The central nervous system, which is comprised of the brain and the spinal cord and the peripheral nervous system, which connects the central nervous system to the body.

c. Gray matter (cell bodies, axon terminals, and dendrites) and white matter (axons).

d. Neurons, which are responsible for transmitting and receiving nerve impulses and glial, which protect neurons and support neurotransmission.

2. The primary excitatory neurotransmitter in the CNS is as follows:
   a. Dopamine
   b. Serotonin
   c. Glutamate
   d. Norepinephrine

3. The brain can be divided into three main regions: the hindbrain, the midbrain, and forebrain. The forebrain contains the cerebrum. The outermost portion of the cerebrum is the cerebral cortex, which contains the occipital lobe, the temporal lobe, the parietal lobe, and the frontal lobe.
   a. True
   b. False

4. The anterior portions of the frontal lobes are called the prefrontal cortex (PFC), which is responsible for executive functions, and includes the dorsolateral prefrontal cortex (DLPFC) and the ventromedial prefrontal cortex (VMPFC). Which of the following sstatements is accurate:
   a. The DLPFC has extensive neural connections to sensory and motor cortices and is involved in the regulation of action, thought, and attention and the VMPFC has extensive connections to the limbic system and is engaged in the regulation of emotional responses.
   b. The VMPFC has extensive neural connections to sensory and motor cortices and is involved in the regulation of action, thought, and attention and the DLPFC has extensive connections to the limbic system and is engaged in the regulation of emotional responses.
   c. Neither the VMPFC has extensive neural connections to sensory and motor cortices and is involved in the regulation of action, thought, and attention nor the DLPFC has extensive connections to the limbic system and is engaged in the regulation of emotional responses.
   d. Only the VMPFC has extensive neural connections to sensory and motor cortices and is involved in the regulation of action, thought, and attention and connections to the limbic system and engagement in the regulation of emotional responses.

Brain Basics

5. The Limbic System:
   a. Is located below the cortical surface; is known as the "logical brain"; is responsible for emotional responsiveness, formation and consolidation of memories, olfaction, and motivation; and is generally thought to include the hypothalamus, the hippocampus, the amygdala, and the cingulate cortex.
   b. Is located below the cortical surface; is known as the "emotional brain"; is responsible for emotional responsiveness, formation and consolidation of memories, olfaction, and motivation; and is generally thought to include the hypothalamus, the hippocampus, the amygdala, and the cingulate cortex.
   c. Is located above the cortical surface; is known as the "emotional brain"; is responsible for emotional responsiveness, formation and consolidation of memories, olfaction, and motivation; and is generally thought to include the hypothalamus, the hippocampus, the amygdala, and the cingulate cortex.
   d. Is known as the "reptilian brain" and is no longer used by humans.

Think About It

1. Science has demonstrated that several variants significantly and adversely impact regions of the brain associated with behavioral and emotional regulation. Should an individual have a functional deficit that compromises behavioral control, should they be held accountable for their actions?
2. In what ways, if any, should knowledge of brain functioning be used to inform criminal justice policies?

Answer Key:
1. (a) 2. (c) 3. (a) 4. (a) 5. (b)

Bibliography

Arnsten, A. (2009). Stress signaling pathways that impair prefrontal cortex structure and function. *Nature Reviews Neuroscience, 10*(6), 410–422. doi:10.1038/nrn2648.

Auer, R., Vittinghoff, E., Yaffe, K., Künzi, A., Kertesz, S., Levine, D., Albanese, E., Whitmer, R., Jacobs, D., Sidney, S., Glymour, M., & Pletcher, M. (2016). Association between lifetime marijuana use and cognitive function in middle age: The coronary artery risk development in young adults (CARDIA) study. *JAMA Internal Medicine, 176*(3), 352–361. doi:10.1001/jamainternmed.2015.7841.

Bronson, J., Stroop, J., Zimmer, S., & Berzofsky, M. (2017). *Drug Use, Dependence, and Abuse Among State Prisoners and Jail Inmates, 2007–2009*. Office of Justice Programs Innovation Partnerships Safer Neighborhoods. Retrieved from www.ojp.usdoj.gov.

Clark, L., & Sahakian, B. J. (2008). Cognitive neuroscience and brain imaging in bipolar disorder. *Dialogues in Clinical Neuroscience, 10*(2), 153–165.

DeLisi, L., Szulc, K., Bertisch, H., Majcher, M., & Brown, K. (2006). Understanding structural brain changes in schizophrenia. *Dialogues in Clinical Neuroscience, 8*(1), 71–78.

Drevets, W. C., Price, J. L., & Furey, M. L. (2008). Brain structural and functional abnormalities in mood disorders: Implications for neurocircuitry models of depression. *Brain Structure and Function, 213*(1–2), 93–118. doi:10.1007/s00429-008-0189.

Giancola, P., & Zeichner A. (1997). The biphasic effects of alcohol on human physical aggression. *Journal of Abnormal Psychology, 106*(4), 598–607.

Hammersley, R., Forsyth, A., Morrson, V., & Davies, J. (1989). The relationship between crime and opioid use. *British Journal of Addiction, 84*, 1029–1043. doi:10.1111/j.1360-0443.1989.tb00786.x.

Harlow, J. (1868). Recovery from the passage of an iron bar through the head. *Publications of the Massachusetts Medical Society, 2*, 327–347.

Howard, R. (2009) The neurobiology of affective dyscontrol: Implications for understanding 'dangerous and severe personality disorder'. In M. McMurran, & R. C. Howard (Eds.), *Personality, personality disorder and violence* (pp. 157–174). Chichester, PA: Wiley.

Kaku, M. (2014). *The future of the mind: The scientific quest to understand, enhance, and empower the mind*. New York: Random House.

Marsh, L., & Krauss, G. L. (2000). Aggression and violence in patients with epilepsy. *Epilepsy Behavior, 1*(3), 160–168.

McDougle, C., Krystal J., Price L., Heninger G., & Charney, D. (1995). Noradrenergic response to acute ethanol administration in healthy subjects: Comparison with intravenous yohimbine. *Psychopharmacology (Berl), 118*(2), 127–135.

Murphy, S., McPherson, S., & Robinson, K. (2014). Non-medical prescription opioid use and violent behaviour among adolescents. *Journal of Child and Adolescent Mental Health, 26*(1), 35–47.

Neylan, T. (1999). Frontal lobe function: Mr. Phineas Gage's famous injury. *The Journal of Neuropsychiatry and Clinical Neurosciences, 11*(2), 280–283.

NIDA. (2015, June 25). Nationwide Trends. Retrieved from www.drugabuse.gov/publications/drugfacts/nationwide-trends.

NIDA. (2017, December 12). Marijuana. Retrieved from www.drugabuse.gov/publications/researchreports/marijuana.

Olsen, R., Hanchar, H., Meera, P., & Wallner, M. (2007). GABAA receptor subtypes: The "one glass of wine" receptors. *Alcohol, 41*(3), 201–209.

Onyike, C., & Diehl-Schmid, J. (2013). The epidemiology of frontotemporal dementia. *International Review of Psychiatry (Abingdon, England), 25*(2), 130–137. doi:10.3109/09540261.2013.776523.

Pearlson, G. (1999). Structural and functional brain changes in bipolar disorder: A selective review. *Schizophrenia Research, 39*(2), 133–140. doi:10.1016/S0920-9964(99)00112-7.

Brain Basics

Schmahl, C., & Bremner, J. (2006). Neuroimaging in borderline personality disorder. *Journal of Psychiatric Research, 40*(5), 419–427. doi:10.1016/j.jpsychires.2005.08.011.

Siddiqui, S. V., Chatterjee, U., Kumar, D., Siddiqui, A., & Goyal, N. (2008). Neuropsychology of prefrontal cortex. *Indian Journal of Psychiatry, 50*(3), 202–208. doi:10.4103/0019-5545.43634.

Vicens, V., Radua, J., Salvador, R., Anguera-Camós, M., Canales-Rodríguez, E., Sarró, S., Maristany, T., McKenna, P., & Pomarol-Clotet, E. (2016). Structural and functional brain changes in delusional disorder. *The British Journal of Psychiatry, 208*(2), 153–159. doi:10.1192/bjp.bp.114.159087.

Volkow, N., Hitzemann R., Wolf A., Logan J., Fowler J., Christman, D., Dewey, S., Schlyer, D., Burr, G., & Vitkun, S. et al. (1990). Acute effects of ethanol on regional brain glucose metabolism and transport. *Psychiatry, 35*(1), 39–48.

Walsh, E., Buchanan, A., & Fahy, T. (2002). Violence and schizophrenia: Examining the evidence *The British Journal of Psychiatry, 180*(6), 490–495. doi:10.1192/bjp.180.6.490.

Wolkin, A., Sanfilipo, M., Wolf, A., Angrist, B., Brodie, J., & Rotrosen J. (1992). Negative symptoms and hpofrontality in chronic schizophrenia. *Archives of General Psychiatry, 49*(12), 959–965. doi:10.1001/archpsyc.1992.01820120047007.

Overview of Advances in Neuroimaging

3

Brain imaging gives us the hope of opening up the black box.
Brian Knutson

If Descartes were alive today, he would be in charge of the CAT and PET scan machines in a major research hospital.
Richard Watson

Learning Objectives

1. Describe the differences between structural and functional neuroimaging.
2. List the common structural neuroimaging and functional neuroimaging techniques.
3. Analyze the potential and limitations of neuroimaging to assist the trier of fact in rendering decisions in the criminal justice process.

Introduction

Assessing the relationship between a subject's brain functioning and criminal culpability has been an accepted aspect of the judicial process for decades, reaffirmed under both the Frye and Dalbert standards of evidence (detailed in Chapter 6). Qualified mental health professionals utilize standardized Forensic Assessment Instruments (FAIs) or Forensic Mental Health Assessments (FMHAs) to assist the trier of fact at all stages of the process, including capacity to waive Miranda, competency to stand trial, criminal responsibility, sentencing and parole, and competency to be executed.

Compromised functioning in a particular region of the brain has historically been inferred by correlating localized brain damage from tumors, head injuries, disease, or pre- or postmortem surgery with changes in personality and behavior.

Now these changes can be captured in images.

In the 20th and 21st centuries, advances in neuroimaging allow jurors and judges to literally glimpse into the mind itself.

64 Neurocriminology

BOX 3.1 COMMON FORENSIC NEUROPSYCHOLOGICAL AND PSYCHOLOGICAL ASSESSMENTS

For decades, the courts have relied upon qualified mental health experts to inform a broad continuum of legal questions. A sample of commonly used tools includes those below. It should be noted that most psychologists use multiple assessments, depending upon the specifics of the referral question.

Capacity to Waive Miranda Rights	Georgia Court Competency Test (GCCT)
	Instruments for Assessing Understanding and Appreciation of Miranda Rights
	Mental Status Exam (MSE)
	Minnesota Multiphasic Personality Inventory (MMPI)
	Peabody Individual Achievement Test (PIAT)
	Test of Memory Malingering (TOMM)
	Structured Interview of Reported Symptoms (SIRS)
	Wechsler Adult Intelligence Scale (WAIS)
	Wechsler Individual Achievement Test (WIAT)
	Wechsler Intelligence Scale for Children (WISC)
	Wechsler Memory Scales (WMS)
	Woodcock Johnson Tests of Cognitive Abilities (WJ)
	Wide Range Achievement Test (WRAT)
	Word Memory Test (WMT)
Competency to stand trial	MacArthur Competence Assessment Tool–Criminal Adjudication (MacCAT-CA)
	Evaluation of Competency to Stand Trial-Revised (ECSTR)
	Competence Assessment for Standing Trial for Defendants with Mental Retardation (CAST-MR)
Criminal responsibility	Hare Psychopathy Checklist–Revised (PCL-R)
	Hare PCL: Youth Version (PCL:YV)
	Mental Status Exam (MSE)
	Minnesota Multiphasic Personality Inventory (MMPI)
	Peabody Individual Achievement Test (PIAT)
	Test of Memory Malingering (TOMM)
	Trail Making Test (TMT)
	Rogers Criminal Responsibility Assessment Scales Structured Interview of Reported Symptoms (SIRS)
	Wechsler Adult Intelligence Scale (WAIS)
	Wechsler Individual Achievement Test (WIAT)
	Wechsler Intelligence Scale for Children (WISC)
	Wechsler Memory Scales (WMS)
	Woodcock Johnson Tests of Cognitive Abilities (WJ)
	Wide Range Achievement Test (WRAT)
	Word Memory Test (WMT)

(Continued)

Overview of Advances in Neuroimaging 65

> **BOX 3.1 COMMON FORENSIC NEUROPSYCHOLOGICAL AND PSYCHOLOGICAL ASSESSMENTS (*Continued*)**
>
> | | |
> | --- | --- |
> | Risk assessment | Hare Psychopathy Checklist–Revised (PCL-R) |
> | | Hare PCL: Youth Version (PCL:YV) |
> | | HCR-20v3 |
> | | Structured Interview of Reported Symptoms (SIRS) |
> | | Violence Risk Appraisal Guide (VRAG) |
> | | Static-99 |
> | | Sexual Violence Risk-20 (SVR-20) |
> | | Minnesota Sex Offender Screening Tool–Revised (MnSOST-R) |
> | | Sex Offender Risk Appraisal Guide (SORAG) |
> | Competence for execution | Mental Status Exam (MSE) |
> | | Minnesota Multiphasic Personality Inventory (MMPI) |
> | | Peabody Individual Achievement Test (PIAT) |
> | | Test of Memory Malingering (TOMM) |
> | | Trail Making Test (TMT) |
> | | Structured Interview of Reported Symptoms (SIRS) |
> | | Wechsler Adult Intelligence Scale (WAIS) |
> | | Wechsler Individual Achievement Test (WIAT) |
> | | Wechsler Memory Scales (WMS) |
> | | Woodcock Johnson Tests of Cognitive Abilities (WJ) |
> | | Word Memory Test (WMT) |

How Neuroimaging Works

Concurrent with the increasing body of knowledge related to brain regions and associated functionality, the last several decades heralded technological advances that enable the less-invasive measurement of these functions and processes. Normative subjects can now be safely compared to those with brain abnormalities. The capacity to image those with known or suspected brain dysfunction, along with control subjects, allows for the more valid and reliable statistical analysis of differences in brain functioning and correlated reactions, responses, perceptions, personality traits, and behaviors—including criminality. This ability underlies the field of neurocriminology.

Neuroimaging can be categorized as structural and functional. Each utilizes tomography—X-ray photographs of a specific plane of the brain—to identify structures or processes that might be associated with pathology. The images are typically displayed in three planes: The axial plane (transverse view) shows an image as if parallel to the ground. The sagittal plane (lateral view) is perpendicular to the axial plane and separates the left and right sides. The coronal plane (frontal view) is perpendicular to the sagittal plane and separates the front from the back (Photo 3.1).

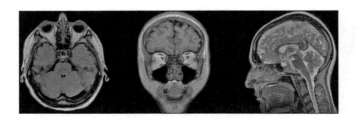

Photo 3.1

Structural Neuroimaging Techniques

Structural neuroimaging techniques allow for the visualization of the brain's anatomical structure by showing contrast between cerebrospinal fluid, white matter (myelinated axons), and gray matter (cell bodies). These techniques allow for the diagnosis of gross (large-scale) pathologies, including those caused by stroke and brain injury. The most common structural imaging techniques are X-ray computed tomography (CT) and magnetic resonance imaging (MRI).

Computed Tomography/Computerized Axial Tomography

In 1972, Allan McLeod Cormack and Godfrey Newbold Hounsfield independently introduced CT, also known as computerized axial tomography (CAT). In contrast to standard X-rays, which allow researchers to look through the body, the CT or CAT scan produces a cross-sectional image that provides the ability to look from within the body. During the CT scan, the subject lies on a table that slides in and out of a hollow, cylindrical apparatus. An X-ray source is affixed to a ring on the inside of the tube, with its beam aimed at the subject's head. The technology produces an X-ray image of a single "slice" of the brain as thin as 1–10 millimeters that can be viewed independently, as well as combined with data from several X-rays, to produce a three-dimensional image—a process that is frequently compared to being able to examine a slice of bread as well as the entire loaf.

Magnetic Resonance Imaging

MRI was developed by researchers Peter Mansfield and Paul Lauterbur, who were subsequently awarded the Nobel Prize in 2003. MRIs produce high-quality and high-resolution two- or three-dimensional images of brain structures without exposing subjects to radiation. During the scan, the subject lies on a movable bed, which is then inserted into the cylinder of the MRI scanner. The MRI machine's magnetism causes

Overview of Advances in Neuroimaging

a realignment of protons found in brain tissue. A radio pulse from the machine causes these protons to move at a specific frequency. The MRI measures the energy that is released when the pulse ceases and the protons return to normal movement. MRI provides detailed structural information and enables the naked eye to distinguish gray matter from white matter. Anatomical techniques, such as diffusion tensor imaging (DTI), have been developed to better visualize myelinated tracts and to assess axonal integrity. DTI uses a measurement known as fractional anisotropy, which indicates the orientation of diffusion along neuronal pathways, and has been applied to evaluate structural integrity following stroke or injury, as well to track the progression of disease states, including Alzheimer's and Schizophrenia.

Neuroscientists have used structural neuroimaging techniques to demonstrate correlations between violence and temporal lobe abnormalities; between reduced prefrontal gray matter volume and antisocial personality disorder and psychopathy; and between structural deficits of the right amygdala and pedophilia (Photo 3.2).

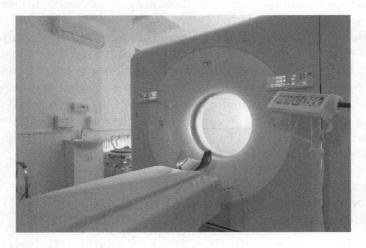

Photo 3.2

Functional Neuroimaging Techniques

The brain utilizes approximately 20% of the oxygen consumed by the body, and requires an uninterrupted oxidative metabolism for functional and structural integrity. Since the brain stores very little oxygen, cerebral brain flow (CBF) must continuously supply needed oxygen. The brain's limited, independent store of oxygen means that a complete interruption in cerebral

68   Neurocriminology

blood flow results in loss of consciousness in less than 10 seconds, and causes irreversible changes in the brain within minutes. Numerous physiological mechanisms work to maintain appropriate CBF, including through the initiation of changes in blood flow and glucose consumption to brain structures engaged in functional activities.

Functional neuroimaging measures neuronal activity by assessing supplies of oxygen and glucose provided by cerebral blood flow (CBF), which increase when subjects complete tasks, such as solving complex problems, or respond to stimuli, as when viewing provocative photographs.

Functional neuroimaging typically measures activity along a color-coded spectrum with red signifying high levels of activity, yellow and green lower levels of activity, and blue the lowest level of activity.

There are several functional brain imaging methods. Generally, these methods are classified into two types: direct measurement of electrophysiological-based activity and indirect measurement of blood flow and metabolic-based activity. Electrophysiological techniques are typically associated with better temporal resolution (i.e., the speed with which image detail can be captured), whereas metabolic-based imaging methods are generally associated with better spatial resolution (i.e., image clarity).

Although there are a range of functional imaging techniques used in a variety of settings, including those dedicated to medical research and clinical intervention, a large-scale study conducted by law professor Deborah Denno found four to be prevalent in criminal justice procedures: Electroencephalography/ evoked potentials (EEG) and magnetoencephalography/magnetic source imaging (MEG), functional magnetic resonance imaging (fMRI), positron emission tomography (PET), and single-photon emission computerized tomography (SPECT).

Electroencephalography/Evoked Potentials and Magnetoencephalography/Magnetic Source Imaging

In 1929, German psychiatrist Hans Berger published his EEG recording of a neurosurgery performed on a 17-year-old boy, introducing the first functional brain measurement technique. Still widely used today, EEG involves the attachment of electrodes, which are mounted to an elastic cap, to the scalp. These electrodes are sensitive to activity of pyramidal neurons which constitute roughly 70% of cells in the cerebral cortex and are, therefore, able to provide measurements related to the regions of the brain engaged in higher functions including sensory perception, generation of motor commands, spatial reasoning, conscious thought, and language. EEG measures activity in the cortical gyri and the depths of the sulci. In contrast, MEG, first introduced in the 1960s, utilizes Superconducting Interference Devices (SQUIDS)—a coil

Overview of Advances in Neuroimaging

that is cooled down to superconductivity by liquid helium—to measure the magnetic fields spontaneously produced by the human brain. In contrast to EEG, MEG arrays are set in a helmet-shaped vacuum flask that is placed on the subject's head. Since the magnetic signals emitted by the brain are of low magnitude, shielding from external magnetic signals, including the earth's magnetic field, is necessary; MEG requires the use of a magnetically shielded room (Photo 3.3).

Photo 3.3

Both EEG and MEG measure brain activity during different brain states (e.g., sleeping and waking), as well as following the introduction of stimuli, such as photographs. In general, EEG is able to measure more of the brain's activity, but is less able to localize this activity, whereas MEG measures less brain activity, but is better able to localize. Both techniques, however, are challenged by what is referred to as to the "inverse problem," that is, the challenge of identifying the source of the underlying signal, which could be a great distance from the point at which it is measured, and affected by head shape or orientation. For this reason, researchers use localization algorithms to determine likely signal source. Some researchers use EEG and MEG simultaneously, which offers the advantage of detecting changes in brain activity in millisecond temporal resolution.

More recently, quantitative EEG (qEEG) has been used to augment the analysis of EEG source imaging. qEEG utilizes more surface electrodes than the standard EEG and applies algorithms that include comparatives to normative databases to create multicolor "Brain Maps" that code functioning (Photos 3.4 and 3.5).

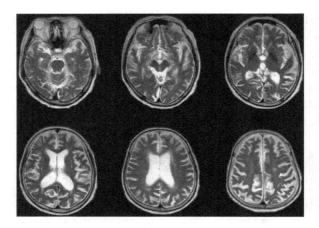

Photo 3.4

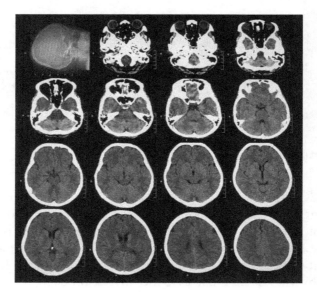

Photo 3.5

Functional Magnetic Resonance Imaging

In 1991, the annual meeting of the Society for Magnetic Resonance in Medicine in San Francisco featured two presentations that introduced the use of functional MR techniques, which employ the same machine that is used for structural MRI scans. In the first presentation, researcher John Belliveau reported the successful use of magnetic resonance to map changes in cerebral blood volume in a subject responding to visual stimuli. A few days later, researcher Thomas Brady and his colleagues presented a video that demonstrated MRI detection of brain activation based on differences in the magnetic properties

Overview of Advances in Neuroimaging

of oxygenated and deoxygenated hemoglobin. When an area of the brain becomes active, there is an increase in neural activity and metabolism and an associated demand for oxygen. Regional cerebral brain flow (rCBF) increases to deliver more oxygenated hemoglobin. The differing magnetic properties of oxygenated and deoxygenated hemoglobin trigger what are called blood oxygen level-dependent (BOLD) measurements, allowing indirect measurement of neuronal activity associated with a particular process.

Other methods of fMRI include arterial spin labeling (ASL), which measures CBF using blood water; blood oxygenation sensitive steady-state (BOSS), which measures a change in signal frequency for deoxyhemoglobin relative to oxyhemoglobin; and dynamic susceptibility contrast (DSC), which requires an injection of gadolinium ions and measures signal disruption around vessels in which it flows, signifying activity.

Positron Emission Tomography

Positron emission tomography (PET) scans integrate CT with radioactive tracers that are injected into the subject and tracked to measure metabolic activities. The tracer enters the brain after approximately 30 seconds, rising to a maximum radiation value approximately 30 seconds thereafter. The decay of the tracer produces particles called positrons, which have an opposing charge to the electrons in the brain. The particles annihilate each other, producing energy in the form of photons. A picture of the rCBF is taken at this time. PET images reflect functional responses within the deep structures of the brain and display changes in brain areas under varying conditions, offering significant sensitivity and specificity.

To measure brain activity, neuroscientists typically use the radioactive tracer compound fluorodeoxyglucose (FDG), a modified glucose molecule. The biggest drawback of PET scanning is that, because the radioactivity decays rapidly, it is limited to monitoring short tasks. PET tracers have been approved to aid in the accurate diagnosis of Alzheimer's disease, which previously could only be done postmortem.

Single-Photon Emission Computerized Tomography

As with PET scans, SPECT imaging uses CT with radioactive tracers. However, the tracer used in SPECT emits gamma rays and stays in the blood stream rather than being absorbed by surrounding tissues, allowing measurement of blood flow variations. SPECT has been specifically employed in the diagnosis of Parkinson's disease.

fMRI research has been used to demonstrate correlations such as between low glucose metabolism in the prefrontal cortex and violence, and low anterior cingulate activity and persistent antisocial behavior (Photos 3.6 and 3.7).

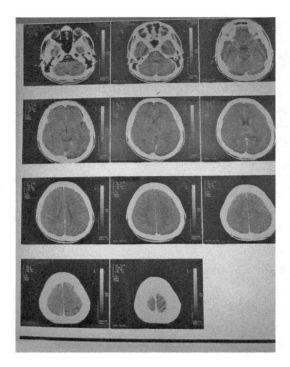

Photo 3.6

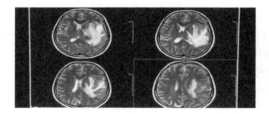

Photo 3.7

Strengths and Limitations of Neuroimaging

Neuroimaging has several significant strengths that have contributed to our understanding of the relationship between brain functioning and human behavior, including criminality.

The relatively noninvasive procedures associated with neuroimaging allow a greater number of subjects to be safely evaluated, providing baseline data from broader segments of the population.

On the individual level, neuroimaging can provide information on the impact of interventions on brain reorganization, which has significant implications for informing crime control and prevention strategies.

Overview of Advances in Neuroimaging

Neuroimaging also has several limitations that have proven particularly relevant when considering its criminal justice applications.

Reliability

Reliability is a measure of internal consistency and stability over time, and one aspect of scientific integrity. It can be established by repeating the same process multiple times and confirming outcome. This is often obtained by having the subject repeat a test after an interval of time has passed, by using different testing tools that measure the same targeted construct, or by having qualified professionals repeat the same experiment—known as inter-rater reliability. Challenges to establishing reliability in neuroimaging include variability in choice of performance tasks, equipment, method of analysis, and even the definitions of the colors used to interpret results. Additional variables that potentially impact reliability are related to the test subject. These variables can include the degree of fidelity to test taking procedures, including remaining completely still, and uncontrollable changes in brain functioning or environmental influences between one examination and another.

Validity

Validity, another measure of scientific integrity, reflects the extent to which an experiment's design measures what it purports to measure.

Neuroimaging can provide a valid measure of the general correlations between brain functioning and criminality by retrospectively imaging those who have engaged in specific crimes and comparing these images against control subjects.

Ironically, the validity of neuroscientific experiments in relation to the individual criminal is compromised by the methodological controls typically employed by researchers to protect the integrity of other experiments: A good experiment attempts to eliminate or minimize the number of confounding variables—distractions in the case of neuroimaging—that might impact the outcome, whereas the criminal is necessarily processing multiple internal and external stimuli at the time of his or her crime. As a result, the neuroimage taken in a simulated experiment with minimal competing demands or stressors might have limited resemblance to the brain's functional response at the time the crime was committed. As it would clearly be unethical to conduct an experiment while a criminal is conducting a crime, it is not possible to reproduce the condition of interest in a laboratory setting.

Inferential Gaps

Neuroimaging measures an individual's brain functioning during a lab-based task. To assert that any identified deficit impacted behavior during an actual, real-world event requires the neuroscientist to apply his or her judgment, and infer a relationship between the deficit and the behavior in question. Such inferences are subject to reasonable dispute and, therefore, cannot definitely answer the categorical questions—such as competent to stand trial or not, able to discern right from wrong—required by the criminal justice system. In short, the neuroimaging examination may lack ecological validity; the stimulus employed during the neuroimaging exam may not trigger the same experience—the affect, the cognition, the literal "state of mind" —of the individual during the crime, rendering it relevance to the trier of fact indirect and subject to debate by competing experts.

Causation

Neuroscience has provided significant data and insight into the association between brain functioning and behavior. Correlation, however, is not causation; the presence of brain dysfunction does not mean that the dysfunction is responsible for the criminal behavior that ensued.

Neuroscience itself has confirmed this by demonstrating the redundancy and cooperation between brain regions and mental processes, and the challenge of disambiguating neural responses to varied stimuli.

In the criminal justice system, this limitation makes it difficult to utilize neuroimaging to respond to the common question of whether the presence of brain dysfunction was a mitigating factor in a specific individual's commitment of a crime.

Group-to-Individual Problem

Although both seek truth, science and the criminal justice system ask two fundamentally different questions: Science asks, "Is this event generalizable to a like sample?" The criminal justice system asks, "Is this event specific to this particular individual?"

This fundamental difference has been called the group-to-individual—or G2i—problem.

Even using large sample sizes, and the most controlled and parallel comparisons, science cannot provide absolutes; conclusions are offered with degrees of confidence that are never 100%. With neuroimaging, these traditional scientific challenges are further complicated by experiments that are typically based upon relatively small sample sizes of less than 25—and in many cases less than 10—and control samples that often are comprised of undergraduate students, who may or may not share the demographic or social characteristics of the sample under study.

The challenge of generalizability in relation to the study of the human brain is further complicated by the variation in individual response; given

Overview of Advances in Neuroimaging

the vast diversity of human experience, and the number of influences that inform it, finding the same response pattern to a particular stimulus in a similar subject group is highly unlikely. It is equally unlikely that the same person will always have the same response.

Neuroimaging has at once opened incredible opportunities to enhance our understanding of human behavior and has reminded that such potential—at least currently—has limitations that must be recognized. In the following chapters, we'll explore the ways in which neuroscience has capitalized on this potential, and navigated these limitations, in both the laboratory and in the criminal justice system.

Key Terms

Tomography: X-ray photographs of a specific plane of the brain.

Axial Plane: Also known as transverse view. A brain image that is parallel to the ground.

Sagittal Plane: Also known as lateral view. A brain image that is perpendicular to the axial plane and separates the left and right sides.

Coronal Plane: Also known as frontal view. A brain image that is perpendicular to the sagittal plane and separates the front from the back.

Structural Neuroimaging: Techniques that allow for the visualization of the brain's anatomical structure by showing contrast between cerebrospinal fluid, white matter (myelinated axons), and gray matter (cell bodies). The most common techniques are X-ray computed tomography (CT) or computerized axial tomography (CAT) and magnetic resonance imaging (MRI).

Functional Neuroimaging: Techniques that measure neuronal activity through direct measurement of electrophysiological-based activity and indirect measurement of blood flow and metabolic-based activity. Includes electroencephalography/evoked potentials (EEG) and magnetoencephalography/magnetic source imaging (MEG), functional magnetic resonance imaging (fMRI), positron emission tomography (PET), and single-photon emission computerized tomography (SPECT).

Use Your Brain

Test Your Knowledge

1. Demonstrating an association between elevated activity in the amygdala and sadism would most likely be identified through the use of:

76   Neurocriminology

   a. Magnetic resonance imaging (MRI)
   b. Functional magnetic resonance imaging (fMRI)
   c. X-ray computed tomography (CT)
   d. Computerized axial tomography (CAT)

2. Demonstrating an association between reduced gray matter volume and antisocial personality disorder would most likely be identified through the use of:
   a. Magnetic resonance imaging (MRI)
   b. Functional magnetic resonance imaging (fMRI)
   c. Positron emission tomography (PET)
   d. Single-photon emission computerized tomography (SPECT)

3. Blood oxygen level-dependent (BOLD) measurements are associated with which neuroimaging technique?
   a. Magnetic resonance imaging (MRI)
   b. Functional magnetic resonance imaging (fMRI)
   c. Positron emission tomography (PET)
   d. Single-photon emission computerized tomography (SPECT)

4. Correlation between two events—such as brain activity and behavior—demonstrates that one event causes the other to occur.
   a. True
   b. False

5. The G2i problem refers to the fundamental principle that:
   a. What is true of a group might not be true for a specific individual.
   b. What is true for one individual 1 day might not be true the next.
   c. What happens in a laboratory does not necessarily translate to the real world.
   d. It is impossible to disambiguate the specific to the general.

Apply Your Knowledge

1. For decades, neuropsychological and psychological assessments have been used by the criminal justice system to assist the trier of fact understand relevant neurological factors that might have influenced a defendant's behavior during the commission of a crime. Should neuroimaging be likewise used? Why or why not?

2. Science and the criminal justice both seek truth, though the approach adopted by each differ. In which areas might the scientific and the criminal justice systems approaches to truth seeking be compatible and mutually beneficial, and in which might ethical or methodological conflicts prove problematic?

Overview of Advances in Neuroimaging

Answer Key:
1. (b) 2. (a) 3. (b) 4. (b) 5. (a)

Bibliography

Greely, H., & Wagner, A. (2011). *Reference manual on scientific evidence* (3rd ed.). Washington, DC: National Academies Press. doi:10.17226/13163.

Grisso, T. (1986). *Evaluating competencies: Forensic assessments and instruments.* New York: Plenum Publishers.

Heilburn, K., DeMatteo, D., Brooks Holliday, S., & LaDuke, C. (2014). *Forensic mental health assessment: A casebook* (2nd ed.). New York: Oxford University Press.

James, O., Doraiswamy, P., & Borges-Neto, S. (2015). PET imaging of tau pathology in Alzheimer's disease and tauopathies. *Frontiers in Neurology 6,* 1–4. doi:10.3389/fneur.2015.00038.

Levy, A. (2006). Bloomberg Markets, Vol. 15, No. 3, February 01, pp. 34–45. Retrieved from www.bloomberg.com/media/markets/neurofinance.pdf.

Raine, A., Buchsbaum, M. S., Stanley, J., Lottenberg, S., Abel, L., & Stoddard, J. (1994). Selective reductions in prefrontal glucose metabolism in murderers. *Biological Psychiatry, 36*(6), 365–366.

Raine, A., Lencz, T., Bihrle, S., LaCasse, L., & Colletti, P. (2000). Reduced prefrontal gray matter volume and reduced autonomic activity in antisocial personality disorder. *Archives of General Psychiatry, 57*(2), 119–127.

Schiltz, K., Witzel, J., Northoff, G., Zierhut, K., Gubka, U., Fellmann, H., Kaufmann, J., Tempelmann, C., Wiebking, C., & Bogerts, B. (2007). Brain pathology in pedophilic offenders: Evidence of volume reduction in the right amygdala and related diencephalic structures. *Archives of General Psychiatry, 64*(6), 737–746. doi:10.1001/archpsyc.64.6.737.

Siegel, G. J., Agranoff, B. W., Albers, R. W., & Molinoff, P. (Eds.). (1999). *Basic neurochemistry: Molecular, cellular and medical aspects* (6th ed.). Philadelphia, PA: Lippincott-Raven.

Watson, R. (2007). *Cogito, ergo sum: The life of René Descartes.* Boston, MA: David R. Godine.

Wong, M. T. H., Lumsden, J., Fenton, G. W., & Fenwick, P. B. C. (1994). Electroencephalography, computed tomography and violence ratings of male patients in a maximum-security mental hospital. *Acta Psychiatrica Scandinavica, 90,* 97–101. doi:10.1111/j.1600-0447.1994.tb01562.x.

Yang, Y., Raine, A., Lencz, T., Bihrle, S., LaCasse, L., & Colletti, P. (2005). Volume reduction in prefrontal gray matter in unsuccessful criminal psychopaths. *Biological Psychiatry, 57*(10), 1103–1108.

Neurocriminology Preliminary Applications

4

I talked to a Doctor once for about two hours and tried to convey to him my fears that I felt come [sic] overwhelming violent impulses. After one session I never saw the Doctor again, and since then I have been fighting my mental turmoil alone, and seemingly to no avail.

The *Whitman Letter*, Sunday, July 31, 1966, 6:45 pm

We weren't lawmakers. We had to give a judgment back the way it was given to us. The evidence being what it was, we were required to send John back insane.

George Blyther, juror at the Hinckley trial

Learning Objectives

1. Assess early cases in which neuroscience was introduced in judicial proceedings to explain criminal behavior.
2. Describe concerns related neuroimaging bias.
3. Consider the potential ramifications of preliminary application of neurocriminology in the criminal justice system to current practices.

Introduction

Society's understanding of aberrant behavior is an important element in creating public policies that are just, that balance the rights of the individual against the needs of the collective, and that ensure reasonable, rational, and appropriate responses.

This is evident in highly publicized crimes, when explanations for criminal behavior are subject to speculation by the public, as well as to investigation and evaluation by the criminal justice system.

Correlations between brain functioning and criminal behavior offer one path to understanding the dynamics involved in engagement in crime. Yet the nuanced information that this science provides has sometimes been met with ambivalence in an environment that seeks definitive answers.

An exploration of some of the highly publicized early cases in which brain functioning was a factor, and the way in which these data were

Charles Joseph Whitman

In the early morning hours of August 1, 1966, Charles Joseph Whitman, a 24-year-old former Marine and University of Texas student, murdered his mother and wife. He then purchased ammunition and a shotgun, murdered two additional people, and climbed the 28 floors of the University of Texas bell tower. Over the course of 96 minutes, Whitman shot 17 people and wounded 31 others before being shot and killed by law enforcement.

Subsequent to his death, authorities found a note in which Whitman stated that he had "been a victim of many unusual and irrational thoughts." He requested that "an autopsy would be performed on me to see if there is any visible disorder."

The medical examiner who performed the autopsy concluded that Whitman died of fatal injuries to the head and to the heart. He also reported an additional finding: "A small brain tumor in the white matter above the brain stem, composed of connective tissue elements of the brain mixed with numerous enlarged blood vessels; no evidence of malignantly fast growth but that of partial tissue death, necrosis." He further concluded that there was no correlation between the tumor and "psychosis or permanent pain," that is, the tumor was not connected to Whitman's behavior.

This conclusion was subsequently questioned by a Commission convened by then Texas governor John Connally. Composed of neurosurgeons, psychiatrists, pathologists, and psychologists, the Commission reexamined paraffin blocks of the tumor and brain tissue, and found features of a gliobastoma multiforme, a highly aggressive and malignant glial tumor that pressed against Whitman's amygdala. The Commission explained the possible and limited inferences that could be derived from this finding:

> While both the physiological and clinical studies are pointing increasingly to certain deeper portions of the brain and the temporal lobe as the substrate for normal and abnormal behavioral patterns involving emotion, the application of existing knowledge of organic brain function does not enable us to explain the action of Whitman on August first.

Consequently, the Commission determined "that the relationship between the brain tumor and Charles J. Whitman's actions on the last day of his life cannot be established with clarity. However, the highly malignant brain tumor conceivably could have contributed to his inability to control his actions."

Neurocriminology: Preliminary Applications

Moreover, the Commission called for "special studies on abnormal behavior":

> This is a dramatic indication of the urgent need for further understanding of brain function related to behavior, and particularly to violent and aggressive behavior. With sufficient knowledge in this area, logical approaches to correction of abnormal behavior can be pursued.

The Commission urged that, "for the public good such studies be supported at a level which would insure rapid progress."

In the half century following the Commission's call, the field of neuroscience made significant progress in this regard. However, the judicial system, scientific community, and the general public have exhibited ambivalence regarding the application of neuroscience within the criminal justice system. Three additional cases in the public domain—those of Jack Ruby, John Hinckley, Jr., and Vincent Gigante—illustrate the distance neuroscience has traveled as well as its limitations in assisting the criminal justice system heed the Whitman Commission's call.

Jack Ruby

On November 24, 1963, 52-year-old Jacob Rubenstein, later known as Jack Ruby, shot and killed Lee Harvey Oswald, the accused assassin of President John Kennedy, as Oswald was being transferred from a city to county jail. The event was televised and replayed over the ensuing days. There was no question that Ruby had committed the crime.

At trial, Ruby's attorney, Melvin Belli, set out to prove that Ruby lacked criminal responsibility for his actions, asserting that Ruby did not know at the time of the murder that his actions were wrong. Specifically, Belli argued that Ruby committed the murder during an epileptic seizure, offering as evidence an abnormality reflected on an on electroencephalogram (EEG) taken while Ruby was in prison. The EEG reflected short bursts of six-per-second activity in the temporal lobes. One defense expert stated that these bursts reflected a "very rare type of epilepsy" associated with a "lack of emotional control, convulsive and excessive types of behavior."

The jury rejected the defense's argument and, on March 14, 1964, convicted Jack Ruby of killing Lee Harvey Oswald, sentencing him to death. The conviction was later overturned on grounds that were unrelated to the scientific evidence; the Appeals Court found that Ruby could not receive a

Neurocriminology: Preliminary Applications 83

fair trial in Dallas in view of the excessive publicity. He died of cancer while awaiting a new trial.

The psychomotor variant identified by the EEG conducted on Ruby is now known to be a normal variant in the general population. Additionally, the link between seizure disorder and criminal activity is more complex than Ruby's defense asserted. Most research has concluded that when violence is committed during the ictal or active stage of a seizure it is typically incidental, rather than volitional and purposeful like the shooting perpetrated by Ruby.

More than illuminate the motivation of an assassin, the evidence related to brain functioning in the Ruby case subsequently served as a caution against the premature introduction of evidence that lacked general acceptance in the scientific community.

One of the next major cases to introduce neuroscience, that of John Hinckley, Jr., likewise demonstrated that such evidence can lack the significance hoped or feared.

John Hinckley, Jr.

On March 29, 1981, John Hinckley, Jr. checked into a Washington, DC, hotel, having arrived by bus after a 1-day stay in Hollywood, California. The following day, after penning a bizarre letter to actress Jodi Foster, Hinckley took a cab to the Washington Hilton hotel, where the President was scheduled to speak at a labor convention. Hinckley carried a loaded 22 caliber revolver. At 2:25, Reagan left the hotel. Hinckley, who was waiting outside, shouted the president's name and, when the president turned, opened fire on him and his entourage from a crouched position. Of the six shots Hinckley fired, one struck Reagan's press secretary James Brady in the head, a second hit policeman Thomas Delahanty in the back of his left shoulder, a third struck Secret Service agent Timothy McCarthy in the chest, and a fourth ricocheted off the president's limousine, hitting him in the chest and lodging in his lung.

The president underwent life-saving surgery. McCarthy recovered and continued to serve an additional 5 years on the presidential protective service detail. Delahanty suffered nerve damage and retired that November. Brady's left forehead and frontal sinus were shattered, and bone fragments as well as metal from the bullet entered his frontal lobe and severed part of the corpus callosum. There was a large blood clot in the right temporal lobe, which distorted and compressed his brain stem. Brady suffered partial paralysis and memory and speech impairment. His death 33 years later, in 2014, was ruled a homicide related to the shooting.

The Secret Service immediately apprehended Hinckley, who was transported to a marine base before transfer to the federal penitentiary in Butner,

North Carolina. The defense argued that Hinckley was not guilty by reason of insanity which, under then federal guidelines, required a finding that "as a result of a mental disease or defect, he lacks substantial capacity either to appreciate the criminality of his conduct or to conform his conduct to the requirements of the law."

Hinckley underwent psychiatric evaluations by experts retained by both the defense and the prosecution. The former included psychiatrist William Carpenter, who testified for 3 days during trial, detailing his conclusion that Hinckley suffered from what he called "process schizophrenia." Carpenter described this disorder as a "severe form of mental illness," that evolved from "fairly subtle disorders" when Hinckley was younger to "more severe disorders and psychotic disorders" at the time of the shootings.

He described Hinckley's symptoms as severe depression and withdrawal from social contacts, eccentric or bizarre thoughts, and delusions, including identity conflation with the Travis Bickle character from the movie *Taxi Driver* in which Foster starred. As evidence for his diagnosis, Carpenter

Neurocriminology: Preliminary Applications

offered findings from his psychological evaluations, including his assessment of Hinckley's behaviors and statements regarding these behaviors. He concluded that Hinckley's primary goal at the time of the shooting was to dramatically suicide in a manner that would somehow unite him with the actress of his obsession.

Notably, when asked if Hinckley lacked the substantial capacity to appreciate the wrongfulness of his conduct, Carpenter offered a nuanced answer:

The ability to reason that is implied in appreciation: I think appreciation of wrongfulness would mean that a person had an ability to reason about it, to think about it, to understand the consequences, to draw inferences about the acts and their meaning. And reasoning processes, which involve both the intellectual component and the emotional component. It is part of what goes together in our reasoning about any issue. That in this regard I believe Mr. Hinckley lacked substantial capacity to appreciate.

The reason for this opinion is that it is an understanding of the very reasoning process he was going through in preparation for and in carrying out the acts, that in his own mind, his own reasoning, the predominant reasoning had to do with two major things, the first of which was the termination of his own existence; the second of which was to accomplish this union with Jodie Foster through death, after life, whatever. But these were the major things that were dominating his reasoning about it. The magnitude of importance to him in weighing and in his reasoning of accomplishing these aims was far greater than the magnitude of the events per se. And in that regard it was not only his mind. He was not able to—he was not reasoning about the legality issue itself.

On the more emotional side of appreciation, which would have to do with some—with the feelings, the emotional appreciation or understanding of the nature of the events, the consequences, he also had an impairment in that regard. And the impairment there was that the emotional consequences of the acts that he conducted were in his experience solely in terms of the inner world he had constructed. The meaning of this to the victims of the act was not on his mind. I don't mean to be crass about this, but in his mental state the effect of this on the President [and] on any other victims was trivial, that they—in his mental state they were bit players who were there in a way to help him to accomplish the two major roles [on] which his reasoning was taking place and were not in and of themselves important to this.

So that I do think that he had a purely intellectual appreciation that it was illegal. Emotionally he could give no weight to that because other factors weighed far heavier in his emotional appreciation.

And as these two things come together in his reasoning process, his reasoning processes were dominated by the inner state—by the inner drives that he was trying to accomplish in terms of the ending of his own life and in terms of the culminating relationship with Jodie Foster.

It was on that basis that I concluded that he did lack substantial capacity to appreciate the wrongfulness of his acts.

Carpenter's assertion that Hinckley's "emotional appreciation" overran his "intellectual appreciation" was objected to by the prosecutor who, in a sidebar, had attempted to limit Carpenter's testimony on appreciation to the more narrow concept of cognition: "to have him go on and talk about emotional appreciation, which is his theory maybe, he then invades the province of the Court. He becomes the definer of the law." The court denied the prosecution's request.

The defense called a second witness, psychiatrist David Bear, to testify as to Hinckley's mental status at the time of the shooting. In addition to his evaluations of Hinckley in the months preceding the trial, Bear based his diagnosis on CAT scans that he conducted. At the time of Bear's testimony, however, the court had not made a final ruling on the prosecution's objection to the scans' admission. Following a bizarre interaction during which Bear was almost held in contempt for refusing to testify unless he was permitted to discuss the CAT scans, he concurred with Carpenter that Hinckley suffered from schizophrenia.

On the day that the defense was prepared to rest its case, the court ruled that the CAT scans were admissible: "Exclusion of this evidence would deprive the jury of a complete picture of the defendant's mental condition."

Bear testified that the scans showed that Hinckley had widened sulci in his brain. He described the findings as powerful evidence of schizophrenia, reporting that approximately one-third of schizophrenics have widened sulci, compared to approximately 2% of the general population.

The trial lasted 2 months. The jury deliberated for 4 days. Hinckley was found not guilty by reason of insanity.

A day after the verdict, an ABC News poll found that 83% of those polled thought "justice was not done" in the Hinckley case. Within a month, Congress conducted hearings on the insanity defense. The resulting passage of the federal Insanity Defense Reform Act of 1984 rendered insanity an affirmative defense, shifting the burden of proof from the prosecution. Additionally, the act required that the defendant prove that a "severe" mental illness made him or her "unable to appreciate the nature and quality of the wrongfulness of his acts"—a far more stringent standard.

In explaining his ruling to admit the CAT scans into evidence, the judge in the Hinckley trial explained that he was influenced, in part, by the defense citation of a prior federal appeals court opinion that concluded that juries should see "all possibly relevant evidence bearing on cognition, volition and capacity."

Failure to include the evidence might provide a valid basis for appeal.

In contrast, the prosecution had leveled concerns that the brain scan would be accorded undue credibility.

Interestingly, jurors who spoke after the verdict did not appear to have given the CAT scan testimony much weight, instead citing "Hinckley's writings" as the basis for determining that he was "a confused guy."

Neurocriminology: Preliminary Applications

The seemingly inconsequential influence of neuroscience to the Hinckley trial notwithstanding, contrasting arguments regarding the potentially probative value versus prejudicial impact of brain scans in the courtroom continues. As illustrated in the case of Vincent Gigante, however, the courts have not unilaterally concluded that this evidence is relevant to issues of cognition, volition, capacity.

BOX 4.1 THE G2i COUNTERARGUMENT

The group to individual—or G2i—challenge in translating neuroscience to the criminal justice system stems from science's focus on general, rather than individual truth, which is necessarily the focus of the courts.

Dr. David Bear, psychiatrist for the defense in the John Hinckley, Jr. trial, offered an interesting argument when challenged on the G2i premise by prosecutor Roger Adelman:

RA: Is it generally the accepted view of all psychiatrists that widened sulci, as they are called, indicate that the person suffering from that phenomenon has schizophrenia? Is that a unanimous view?

DB: It is nobody's view.

RA: Isn't it true that the studies you are talking about indicate that most people who are schizophrenic don't have widened sulci?

DB: To be precise about the word "most": In one study from St. Elizabeth's Hospital, one-third of the schizophrenics had widened sulci. That is a high figure. It is true that the simple majority didn't...but the fact that one-third had these widened sulci—whereas in normals, probably less than one out of 50 have them—that is a very powerful fact.

RA: That is a fact?

DB: Yes. Yes. It is a statistical fact ... It is as much a fact as this: We know statistically that a male who smokes 10–20 packs of cigarettes is 20 times more likely to get lung cancer. That is a fact and it is a strong enough fact for the Surgeon General to write on the package, "Don't smoke." Yet it is very clear nobody can say that anyone who takes the next cigarette will get lung cancer. It is a statistical fact, as I mentioned, that one-third of schizophrenics have widened sulci and probably less than 2% of the normal people have them. That is a powerful statistical fact and it would bear on the opinion in this case.

Vincent Gigante

Beginning in the mid-1960s, Vincent Gigante feigned mental illness by frequently walking in his Greenwich Village, New York neighborhood in his pajamas and mumbling to himself. In 1990, he was charged with murder and labor racketeering; law enforcement charged that Gigante—known by the monikers "the Chin" and "the Oddfather"—was the head of the Genovese crime family. Gigante's defense attorney maintained that his client was incompetent to stand trial due to "dementia rooted in organic brain damage."

The prosecution and the defense called on expert and nonexpert witnesses to testify to Gigante's mental status. Gigante was found competent to stand trial and subsequently convicted of racketeering and conspiracy to murder.

Gigante's attorneys filed a motion that Gigante be declared incompetent to be sentenced. Pursuant to federal law, a defendant can request a hearing prior to sentencing if there is reasonable cause to believe that, based upon a preponderance of the evidence, he is suffering from a mental disease or defect that renders him mentally incompetent to the extent that he is unable to understand the nature and consequences of the proceedings against him or to assist properly in his defense. If found mentally incompetent, Gigante would be hospitalized rather than incarcerated.

Neurocriminology: Preliminary Applications

Among the defense experts called to testify was Dr. Monte S. Buchsbaum, a physician who had also offered PET scan evidence at Gigante's competency to stand trial hearing.

The court noted in the competency to be sentenced ruling:

> His [Dr. Buchsbaum's] studies were previously declared by this court to be inadequate to support his opinion at hearings to determine defendant's competence to stand trial, because of the lack of a sufficient scientific database to support the conclusions he drew from defendant's PET scan. His conclusion "that the abnormal metabolic pattern seen, which is typical of patients with dementia is not the product of psychopharmacological drug action," was not persuasive in the absence of further scientific studies... Even as supported by his report of November 16, 1997, relying on additional studies of doctor van Gorp [sic] and others, Doctor Buchsbaum's opinion is not persuasive. His diagnosis was that "PET scan abnormalities indicated possible dementia." He also testified that defendant's electroencephalogram is "consistent with a person suffering from dementia." At most, the work of Doctor Buchsbaum supports the conclusion that there is some damage to defendant's brain, resulting in some one or the other form of dementia. The critical issue of the balance between malingering and dementia in explaining defendant's actions is not illumined by Doctor Buchsbaum's work.

In short, the court determined that, although neuroscience might indeed substantiate the presence of a brain abnormality, it did not demonstrate that the abnormality caused Gigante's actions, nor did it prove that he was not malingering incompetency.

The court ruled Gigante competent, and he was sentenced to 12 years. In April 2003, Gigante appeared before a federal judge in Brooklyn and pleaded guilty to obstruction of justice.

During his sentencing, Gigante stated that he "knowingly, intentionally misled doctors" who evaluated his mental competency. He was sentenced to an additional 3 years in prison and 3 years of probation.

Gigante died in prison in 2005.

Early cases in the public domain that examined issues of brain dysfunction introduced the general population to neurocriminology, concurrent with the courts' efforts to determine its appropriate role. The cases of Whitman, Ruby, Hinckley, and Gigante are significant for illustrating some of the approaches that the criminal justice system has employed to navigate the potential and limitations of neuroscience.

Neither the courts nor jurors demonstrated the neuroimage bias that had been feared when imaging was first introduced at trial, and that early research suggested might occur—what has been referred to as the "CSI effect."

A series of studies conducted in 2008 found that laypersons deemed information to have greater credibility when neuroimages were presented. Later studies conducted between 2011 and 2013 found no such bias. More recently, research suggests that the issue of neuroimage bias may be too complex to be resolved categorically. Rather, it appears that this bias may occur under certain circumstances. A 2013 study, for example, found that neuroimaging was impactful when introduced after information that did not contain neuroimaging—a finding that suggests that undue credibility could be accorded to neuroimaging evidence presented by the defense if this type of evidence is not first introduced by the prosecution.

In the years since these cases, continued neuroscientific research has expanded the knowledge base related to brain–behavior correlates of criminality. Consequently, it is likely that courts will continue to consider evidence of brain dysfunction as deemed relevant to informing the appropriate disposition of justice.

The next chapter will highlight findings from the growing body of research on the brain functioning of those who have committed crimes.

Key Terms

Insanity Defense Reform Act of 1984: Comprehensive federal legislation enacted after the John Hinckley, Jr. was found not guilty by reason of insanity in the assassination attempt on then President Ronald Reagan. Created more stringent standards including making insanity an affirmative defense and requiring proof that a severe mental illness rendered a defendant unable to appreciate the nature and quality of the wrongfulness of his acts.

Neuroimage Bias: Perception that the credibility of evidence is more significant and relevant based upon the introduction of neuroimaging scans.

Use Your Brain

Test Your Knowledge

1. The special Commission to investigate the mass murder committed by Charles Whitman found that Whitman had a gliobastoma multiforme tumor of the brain. The Commission concluded that this:
   a. Caused Whitman's violent behavior.
   b. Was unrelated to Whitman's violent behavior.
   c. Could have contributed to Whitman's violent behavior.
   d. Contributed to his mental illness but not his violent behavior.
2. The defense in the Jack Ruby case asserted that he committed his assassination of Lee Harvey Oswald while in the ictal stage of epileptic seizure. This is consistent with current research that suggests an association between criminal violence and brain dysfunction evidence during this seizure stage.

Neurocriminology: Preliminary Applications

a. True
b. False

3. Following the jury's verdict of Not Guilty by Reason of Insanity in the John Hinckley trial, Congress passed the Insanity Defense Reform Act of 1984 which changed the legal standard for insanity. Such a claim now requires a defendant to prove that he/she:
   a. Has a mental disease or defeat and lacks substantial capacity either to appreciate the criminality of his/her conduct or to conform to the requirements of the law.
   b. Has a severe mental illness that renders him or her unable to appreciate the nature and quality of the wrongfulness of his acts.
   c. Has a mental disease or defect that results in an irresistible or uncontrollable impulse that prevents compliance with the law.
   d. Has a severe mental illness that prevents him/her from following the law.

4. In the case of Vincent Gigante, the court found neuroscientific evidence of brain abnormality irrelevant to determining Gigante's competency to be sentenced due to:
   a. The inaccuracy of the PET scan.
   b. Conflicting evidence of brain abnormality.
   c. Insufficient research to suggest that the brain abnormality identified in the PET scan was relevant to Gigante's competency.
   d. Concerns that neuroimaging might bias the jury.

5. The most recent research on neuroimage bias suggests that:
   a. Concerns are unfounded
   b. There is generally neuroimage bias
   c. There may be neuroimage bias under certain circumstances
   d. There is always neuroimage bias

Apply Your Knowledge

1. One of the limitations of the translation of neuroscience to the criminal justice system—which was highlighted in each of the public domain cases included in this chapter—is that correlation of brain dysfunction with behavior does not prove that the identified brain dysfunction causes behavior. Given this limitation, how—if at all—do you think that neurocriminology is useful to the criminal justice system?

2. In what context(s) could neuroimage bias arise in relation to neurocriminology? What steps, if any, can be taken to mitigate this bias?

Answer Key:
1. (c) 2. (b) 3. (b) 4. (c) 5. (c)

Bibliography

Baker, D., Ware, J.M., Schweitzer, N.J., & Risko, E.F. (2015). Making sense of research on the neuroimage bias. *Public Understanding of Science, 26*, 1–8. doi:10.1177/0963662515604975.

Caplan, L. (1987). *The insanity defense and the trial of John W. Hinckley Jr.* Miller Place, NY: Laurel.

Cytowic, R. (1981). The Long Ordeal of James Brady. *New York Times.* Sept. 27.

Gurley, J.R., & Marcus, D.K. (2008). The effects of neuroimaging and brain injury on insanity defenses. *Behavioral Sciences & The Law, 26*(1), 85–97. doi:10.1002/bsl.797.

Gutmann, L. (2007). Jack Ruby. *Neurology, 68*(9), 707–708. doi:10.1212/01.wnl.0000256034.72576.13.

Low, P., Jeffries, J., Bonnie, & R. (1986). *The Trial of John W. Hinckley, Jr: A Case Study in the Insanity Defense.* Foundation Press. Retrieved from https://openlibrary.org/publishers/Foundation_Press.

McCabe, D.P., & Castel, A.D. (2008). Seeing is believing: The effect of brain images on judgments of scientific reasoning. *Cognition, 107*(1), 343–352. doi:10.1016/j.cognition.2007.07.017.

McCormally, S. (1982). John W. Hinckley Jr., Found Innocent by Reason of…United PressInternational.June23.Retrievedfromwww.upi.com/Archives/1982/06/23/John-W-Hinckley-Jr-found-innocent-by-reason-of/9581393652800/.

Michael, R.B., Newman, E.J., Vuorre, M., Cumming, G., & Garry, M. (2013). Of the (non)persuasive power of a brain image. *Psychonomic Bulletin & Review, 20*(4), 720–725. doi:10.3758/s13423-013-0391-6.

Raab, S. (1989). Family Asks Judge to Find Mafia Boss Mentally Ill. *New York Times.* Dec. 29.

Raab, S. (1991). Experts Assess Mental Fitness of a Defendant. *New York Times.* March 12.

Report on the Charles J. Whitman catastrophe. (1966). Texas Governor's Committee and Consultants. Archives and Information Services Division, Texas State Library and Archives Commission. Retrieved from https://legacy.lib.utexas.edu/taro/aushc/00489/ahc-00489.html.

Schweitzer, N.J., Baker, D.A., & Risko, E.F. (2013). Fooled by the brain: Re-examining the influence of neuroimages. *Cognition, 129*(3), 501–511. doi:10.1016/j.cognition.2013.08.009.

Schweitzer, N.J., Saks, M.J., Murphy, E.R., Roskies, A.L., Sinott-Armstrong, W., & Gaudet, L.M. (2011). Neuroimages as evidence in a Mens Rea Defense: No impact. *Psychology, Public Policy, and Law, 17*(3), 357–393. doi:10.1037/a0023581.

Slade, M., & Biddle, W. (1982). The State of Hinckley's Brain. *New York Times.* June 6.

Taylor, S. (1982). Shootings by Hinckley Laid to Schizophrenia. *New York Times.* May 15.

Taylor, S. (1982). Judge Rebukes Hinckley Witness over CAT Scan. *New York Times.* May 20.

United States v. Gigante, 982 F. Supp. 140 (E.D.N.Y. 1997).

Criminals in the Lab

5

Even though neuroscience, or any science examining the causes of and contributions to human behavior, can help reveal fallacies in our previously held normative judgments, neuroscience does not inevitably lead us to the answers to fundamental moral questions.

Presidential Commission for the Study of Bioethical Issues, March 2015

Neurons giveth and neurons taketh away.

Abhijit Naskar

Learning Objectives

1. Describe the experimental design generally used to research brain dysfunction and criminal behavior.
2. Assess the known correlates between brain functioning and crime subtypes.
3. Analyze the opportunities and challenges inherent in translating neuroscientific findings from the laboratory to the criminal justice system.

Introduction

Through technology that noninvasively evaluates the brain, neuroimaging allows scientists to compare brain functioning within and between groups who have engaged in selected crimes, as well as between criminals and noncriminal control groups. Available research typically involves imaging on a small subset of individuals who have engaged in a particular criminal behavior, meta-analysis of studies related to the target behavior, or case studies, particularly in relation to subjects whose criminality is correlated with acquired dysfunctions.

The group to be studied is compared to a control group that is matched to the greatest extent possible on conditions that would otherwise represent confounding variables, including age, IQ, gender, ethnicity, social class, dominant handedness, substance abuse history, and mental health history. Those with claustrophobia or metal implants are screened out.

93

BOX 5.1 CONTINUOUS PERFORMANCE TESTS

One of the common stimuli used by neuroscientists to assess brain functioning is the Continuous Performance Test (CPT).

The first CPT was developed by psychologists Haldor Rosvold, Allan Mirsky, Irwin Sarason, Edwin Bransom, and Lloyd Beck in 1956 to study vigilance. The test required subjects to respond whenever the letter "X" appeared in a series of rapidly presented visual materials. The more complete a subject's responses, the higher his or her vigilance or attention was rated.

Subsequent variations of the original CPT utilize letters, numbers, or audio cues presented at varied intervals. Typically, the subject must press a button or computer mouse each time the designated stimulus is presented.

In addition to calculating correct responses as a measure of sustained attention, researchers also calculate the number of incorrect responses, which correlate with impulsivity. A pattern of random responding has been associated with dyscontrol.

Engagement in CPTs has been associated with elevated glucose metabolic rates, which allows for the measurement of brain activity during functional imaging studies.

In structural imaging studies, the brain volume of those in the study group is measured against those in the control group to identify potential statistical differences. In functional imaging research, such differences are likewise explored, generally by having each group engage in an activity such as responding to suggestive and neutral stimuli, or engaging in a continuous performance test (CPT). In normal controls, CPT engagement generally results in higher glucose activity in the frontal, temporal, and parietal lobes (see Box 5.1).

Over the past several decades, these efforts have shed light on some of the shared and distinguishing structural and functional attributions of criminal subtypes.

Murderers

In 1994, psychologist Adrian Raine first conducted a brain imaging study on murderers. Raine and his colleagues used positron emission tomography (PET) scans to evaluate metabolic activity in 41 individuals who had been charged with murder or manslaughter and were seeking evidence related not to guilty by reason of insanity (NGRI) pleas.

Criminals in the Lab

The 41 were age and sex-matched to control subjects. The murderers in Raine's study showed decreased glucose activity in the frontal lobe (bilateral prefrontal cortex), which is associated with executive functioning and the ability to regulate impulsivity and aggression, and the parietal lobes (posterior parietal cortex), associated with reading and arithmetic ability and moral decision-making. The murderers also showed decreased activity in the corpus callosum, which may result in compromised processing of negative emotions in the right hemisphere by the left. Of additional interest, the murderers were found to have abnormal asymmetries in glucose activity in the amydala, thalamus, and medial temporal gyrus. The consistently relative reduced left activity and relative greater right activity is associated with possible compromises in appraisal and appropriate responding to the social environment.

Predatory versus Affective Murderers

Two distinct types of violence in which murderers engage are instrumental violence, characterized by planning designed to accomplish a desired goal, and impulsive/reactive violence, which are unplanned and uncontrolled acts typically triggered by real or perceived insults or threats.

Forensic psychologist Reid Meloy has labeled these distinctions as "predatory versus affective violence."

In 1998, Raine and colleagues published findings from research that explored differences in the brain functioning of affective and predatory murderers. Using the PET scans that Raine obtained on the 41 murderers from his prior research, two independent researchers reviewed relevant documents that included the murderers' criminal records, psychological assessments, medical records, and court transcripts. The researchers rated the murderers on a 4-point predatory-affective violence scale. Only the 15 who were rated as "1" (strongly predatory) or "4" (strongly affective) were included in the study.

Both affective and predatory murderers were found to have significantly higher levels of right subcortical glucose metabolism. Affective murderers, however, were found to have lower prefrontal activity. In contrast, predatory murderers were found to have prefrontal activity that was consistent with the non-murder control subjects.

With the highly active "bottom–up" engagement of the subcortical limbic area and compromised "top–down" executive functioning of the prefrontal lobe, one would expect to see dysregulated emotion that is unchecked and susceptible to misinterpretation of environmental cues. This finding is consistent with the affective murderer's engagement in emotionally laden violence.

96 Neurocriminology

Similarly, the findings related to predatory murderers suggested that they could regulate and direct their aggression by calling upon the planning and sequencing abilities afforded by functional prefrontal lobes.

**BOX 5.2   THE BRAIN AND JUSTIFIED
VERSUS UNJUSTIFIED HOMICIDE**

Most crimes under U.S. law require the defendant to possess mens rea or "guilty mind." Specifically, the justice system generally differentiates between those who know that their actions are wrong, will cause harm to another, and intentionally proceed, and those who lack such knowledge or the capacity for such knowledge. Classic distinguishing examples include those who kill another for vengeance versus for self-defense, mothers who commit infanticide during an episode of post partum psychosis versus to be rid of an unwanted child, or a driver who intentionally runs down a pedestrian versus one who is unable to brake in time.

Mens rea is associated with the concepts of moral and legal culpability. In the examples above, all parties could be considered morally culpable, but only those who knowingly and intentionally engaged in their acts would be considered criminally culpable. In effect, one with a guilty mind knows that he or she will cause harm and does it anyway.

Is guilt associated with identifiable changes in brain functioning?

Researchers Molenberghs, Ogilvie, and colleagues sought to find out through an experiment designed to determine if the brain responds differently when confronted with justified versus non-justified killing. The researchers recruited 48 male and female participants to participate in an fMRI study during which they were to imagine that they were responsible for the acts that occurred in a first-person perspective of video game clips they viewed. The clips portrayed a soldier, a civilian, or nobody being shot. The use of video viewing controlled for differences that would otherwise be introduced if the subjects actually played the games during the experiment. After viewing, the respondents indicated who they shot by pressing a button on a response pad.

Results from the fMRIs found that participants demonstrated significantly greater activation in the lateral orbitofrontal cortex (OFC) when imagining shooting civilians versus shooting soldiers. Correlating the imaging results with a post-fMRI questionnaire, the researchers found that the higher the participant's reported level of guilt, the greater the lateral OFC activity. They concluded that violence that is seen as justified can lead to less activation of the brain responses typically associated with harming another.

(Continued)

Criminals in the Lab 97

> **BOX 5.2   THE BRAIN AND JUSTIFIED VERSUS UNJUSTIFIED HOMICIDE (*Continued*)**
>
> A significant body of research has demonstrated that the OFC is involved in executive control through the inhibition of neural activity associated with irrelevant, unwanted, or painful stimuli. Additionally, the lateral OFC has been implicated in efforts to suppress emotional responses in situations requiring social judgment.
>
> In other words, the researchers' findings suggest that mens rea may be associated with heightened lateral OFC activity.

Murderers with Severe Mental Illness

The vast majority of those who are mentally ill do not engage in crime. In addition to underscoring the complex interplay of non-neurological factors associated with criminality, this finding speaks to the challenge of isolating the neurobiological bases of violent behavior. Neuroimaging has contributed to an understanding of the variants previously correlated with both extreme violence and some forms of mental illness, including schizophrenia.

Raine and colleagues including research psychologist Yaling Yang advanced knowledge on the potential distinction between the processes associated with violence and those implicated in mental illness through a 2010 study on schizophrenia. Specifically, the researchers used MRIs to compare the brains of 22 murderers with schizophrenia, 18 murderers without schizophrenia, 19 nonviolent patients with schizophrenia, and 33 normal controls. Murderers were detainees accused of homicide who were undergoing forensic psychiatric evaluations, whereas the nonindividuals with schizophrenia were hospital inpatients. Normal controls were community members without history of mental illness.

As in other research related to murderers, the Raine–Yang study also implicated the limbic system in individuals with schizophrenia who committed murder. Consistent with prior studies on mental illness, the researchers also found that individuals with schizophrenia had reduced gray matter volume in the dorsolateral prefrontal cortex when compared to those without schizophrenia. However, the researchers also found a distinction between murderers with the disease and non-murderers: murderers with schizophrenia showed significant gray matter volume reductions in the right hippocampus, as well as the surrounding parahippocampal gyrus, when compared to both nonviolent patients with schizophrenia and nonschizophrenic control subjects.

Consistent with brain–behavior correlates associated with other murder subtypes, compromises to the right subcortical region can result in emotional

Adolescent Murderers

Adolescents who commit homicide have also been shown to have reduced gray matter volumes in the temporal lobes generally, and the hippocampus and parahippocampal area specifically. A 2014 study by Cope and colleagues, for example, compared adolescent murderers against both non-homicidal incarcerated adolescents and community controls. Of particular significance was the researchers matching of the homicidal and non-homicidal adolescents on key demographic and psychometric factors, including the PCL-YV, which measures psychopathy. Based upon the controls that they also performed for age, IQ, socioeconomic status, traumatic brain injury, mental disorder, conviction rates, and substance abuse, the researchers found that the difference in gray matter volume was the only differentiator between homicidal and non-homicidal adolescents. Interestingly, the researchers found that the gray matter volume reductions in youth who committed homicides were bilateral. This contrasts with findings from research on adult murderers, which have shown significant differences in the right hippocampal and parahippocampal regions. These differences may be attributed to brain development; the peak of hippocampal gray matter development has been estimated to occur from 11 years to after 30 years, depending on research methodology, intervening environmental factors, and genetic heterogeneity.

Coupled with other studies that implicate this region, the Cope's study further suggests an association between compromises in the hippocampal and parahippocampal regions and homicidal violence (Photo 5.1).

Non-Homicidal Offenders

Non-homicidal offenders are a heterogeneous group, and studies have defined their behavior along a continuum from aggression to criminal battery. Research on these subpopulations includes studies on acquired antisocial behavior, such as following traumatic brain injury or lesion development (although a predisposing genetic condition may also be present), and retrospective analyses of those who have engaged in non-homicidal violence, such as intimate partner violence.

In 2016, for example, Bueso-Izquierdo and colleagues published an fMRI study that involved 21 batterers and 20 other criminals convicted in Granada, Spain. Each subject observed a series of images that depicted intimate partner violence and general violence, as well as neutral images. When shown images depicting general violence and intimate partner violence, those who had engaged in intimate partner violence showed higher activation in the OFC, associated

Criminals in the Lab

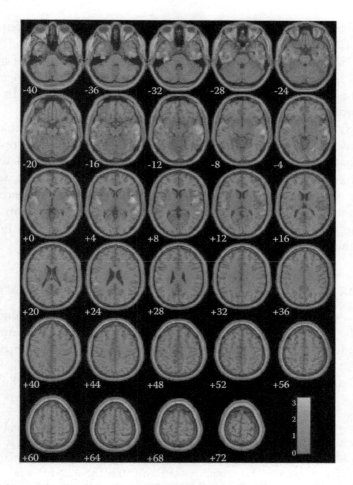

Photo 5.1 Cope and colleagues used structural MRI scans to assess regional gray matter volume differences between adolescent homicide offenders (*n* = 20) versus non-homicide offenders (*n* = 135). These results suggest that after controlling for important moderating variables, youth homicide offenders show the greatest gray matter deficits in bilateral paralimbic regions including the medial and lateral temporal lobes, particularly the hippocampus, posterior insula, superior temporal gyrus, middle temporal gyrus, parahippocampal gyrus, fusiform gyrus, and inferior temporal gyru.

with the processing of social cues and decision-making. Additionally, whereas the general criminal group did not exhibit different responses to general violence versus intimate partner violence images, the batterers did; when viewing images depicting intimate partner violence, batterers exhibited activation in the medial prefrontal cortex, posterior cingulate cortex, and the left angular gyrus. Given the cingulate cortex's involvement in the formation of long-term memories for emotionally significant events, it was hypothesized that batterers

100   Neurocriminology

may experience increased negative feelings that raise fear of abandonment and increase risk of maladaptive behavior, including violence.

In 2015, Bannon and colleagues conducted a synthesis study of structural MRI studies conducted on violence associated with a range of etiologies including head injuries, traumatic brain injuries, intracranial injuries, lesions, and frontal lobe injuries.

The synthesis study broadly implicated the frontotemporolimbic regions of the brain. Consistent with findings from the batterers' study, the researchers found that non-homicidal violence was particularly correlated with gray matter volume reductions in the OFC. These reductions, in turn, were posited to impact neural communication with the interconnected anterior cingulate, and amygdala, compromising problem solving, perception of social cues, and the ability to regulate negative emotions, and in turn increasing risk for maladaptive violent responses.

BOX 5.3   CASE STUDY: FRONTOTEMPORAL DEMENTIA

Consistent with findings related to those who engage in non-homicidal aggression and who have been found to have similar brain dysfunction, individuals with frontotemporal dementia can exhibit inappropriate and antisocial behaviors.

Drs. Serggio Lanata and Bruce Miller of the University of California's Memory and Aging Center offer the following case study:

The patient was 55 years of age when she first came to our attention. She was right-handed. Her doctor referred her to our centre at the insistence of the patient's husband, who upon learning of bvFTD [behavioral variant Frontotemporal Dementia] thought his wife may have the disease. The first serious "red flag" occurred during a family trip abroad 4 years prior to presentation; she made several racial comments about a restaurant server despite clearly causing discomfort to everyone present, including the server. A few months later, she repeatedly interrupted her son's graduation ceremony by initiating conversation with guests during the ceremony. She also attempted to take a bouquet of flowers from another family to give it to her son. Her behaviour in social situations became increasingly erratic over the ensuing 2 years. She regularly initiated conversations with strangers about sex, and also made sexual jokes in front of children. She worked in an elementary school at the time, and her supervisor noted that she developed a tendency to play roughly with children, to the extent that she made some of them cry. She lost her job 3 years after onset of her illness. One year before seeing us, she developed a voracious appetite for sweets leading to a 25-pound weight gain within 12 months. By the time we

(Continued)

BOX 5.3   CASE STUDY: FRONTOTEMPORAL DEMENTIA (*Continued*)

first saw her in our centre, her behaviours had diminished in intensity. Her neurological examination was notable for stimulus-bound behaviours and lack of respect for social boundaries. She repeatedly tried to kiss and hug the examiner. There were no signs of MND [motor neuron disease]. Neuropsychological testing was notable for a score of 26/30 on the mini-mental state examination. She was presented with 16 photographs of faces with a particular emotion, and was able to correctly identify the emotion of only 6 faces (normal = 13.3 ± 1.7). On the Interpersonal Reactivity Index–Perspective Taking (IRI-PS), she scored 8/35 (normal = 24.5 ± 5.5), suggestive of a profound inability to think of others' emotions; likewise, she scored 15/35 (normal = 28.6 ± 4.5) on the IRI Empathic Concern (IRI-EC) scale, suggestive of a profound impairment in her ability to experience the emotional state of others. Her brain MRI revealed marked bilateral frontal and anterior temporal lobar atrophy. Genetic testing revealed a C9ORF72 gene expansion (interestingly, the patient's family history was only notable for a distant paternal uncle with suspected Parkinson's disease). Her syndrome evolved into a predominantly apathetic state.

She died at age 59 years, approximately 8 years after onset. Brain autopsy confirmed FTLD.

Sexual Offenders

Imaging studies related to adult sexuality have found that viewing sexually explicit material is associated with subcortical activation of the hypothalamus, thalamus, and caudate nucleus, and cortical activation in the dorsolaterial and dorsomedial prefrontal cortex regions.

Research on sexual offenders, including rapists and pedophiles, has demonstrated different—and opposing—response patterns in these subgroups compared to normal controls.

Rapists

In a 2016 study of 15 male rapists of female stranger victims, Chen, Raine, and colleagues used diffusion tensor imaging (DTI) to identify potential white matter abnormalities. The researchers found significant fractional anisotropy (FA) increases near the angular gyrus, posterior cingulate, and the medial frontal pole of the rapist group versus a matched control group. These areas are associated with moral decision-making. The posterior cingulate is also interconnected with the parahippocampal gyrus area that has been implicated in studies of psychopaths and individuals who have committed murder. The researchers also found heightened activity in areas associated with general male sexual responsiveness, which could indicate a tendency toward overarousal states (Photo 5.2).

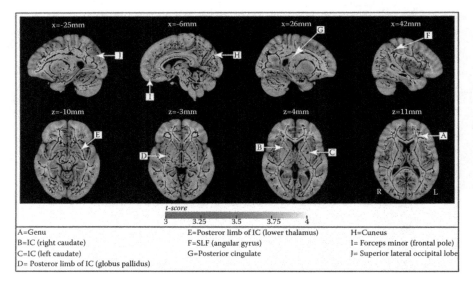

Photo 5.2 Using diffusion tensor imaging (DTI), Chen, Raine, and colleagues identified significant fractional anisotropy (FA) increases near the angular gyrus, posterior cingulate, and the medial frontal pole in Arapist subject group.

> **BOX 5.4   CASE STUDY: PEDOPHILIA AND RIGHT ORBITOFRONTAL TUMOR**
>
> Research has demonstrated that the brains of pedophiles do not respond to adult erotic materials in the same manner as non-pedophiles. It would clearly be unethical to conduct an experiment that evaluated brain functioning when subjects view child pornography. Inferential correlations between the right orbitofrontal region and pedophilia, however, have been made based on case studies. One of the most compelling case studies was published by Drs. Jeffrey Burns and Russell Swerdlow of the University of Virginia Health System:
>
> > A 40-year-old, right-handed man in an otherwise normal state of health developed an increasing interest in pornography, including child pornography. He had a preexisting strong interest in pornography dating back to adolescence, although he denied a previous attraction to
>
> *(Continued)*

Criminals in the Lab

BOX 5.4 CASE STUDY: PEDOPHILIA AND RIGHT ORBITOFRONTAL TUMOR (*Continued*)

children and had never experienced related social or marital problems as a consequence. Throughout the year 2000, he acquired an expanding collection of pornographic magazines and increasingly frequented Internet pornography sites. Much of this prurient material emphasized children and adolescents and was specifically targeted to purveyors of child pornography. He also solicited prostitution at "massage parlors," which he had not previously done.

The patient went to great lengths to conceal his activities because he felt that they were unacceptable. However, he continued to act on his sexual impulses, stating that "the pleasure principle overrode" his urge restraint. He began making subtle sexual advances toward his prepubescent stepdaughter, which he was able to conceal from his wife for several weeks. Only after the stepdaughter informed the wife of the patient's behavior did she discover with further investigation his emerging preoccupation with pornography, and child pornography in particular. The patient was legally removed from the home, diagnosed as having pedophilia, and prescribed medroxyprogesterone [Depo-Provera]. He was found guilty of child molestation and was ordered by a judge to either undergo inpatient rehabilitation in a 12-step program for sexual addiction or go to jail. Despite his strong desire to avoid prison, he could not restrain himself from soliciting sexual favors from staff and other clients at the rehabilitation center and was expelled. The evening before his prison sentencing, he came to the University of Virginia Hospital (Charlottesville) emergency department complaining of a headache. A nonphysiologic cause was suspected, and the psychiatry service admitted him with a diagnosis of pedophilia, not otherwise specified, after he expressed suicidal ideation and a fear that he would rape his landlady. The day after his admission he complained of balance problems, and a neurologic consultation was obtained.

The patient's medical history was notable for a closed head injury 16 years earlier that was associated with a 2-minute loss of consciousness and no apparent neurological sequelae, a 2-year history of migraines, and hypertension. He was without a previous psychiatric or developmental history and had exhibited no prior deviant sexual behavior. Medications included fluoxetine hydrochloride, amlodipine besylate, metoclopramide hydrochloride (for nausea), and medroxyprogesterone acetate at a dose of 10 mg/d. There was no family history of psychiatric disease. He had worked as a corrections officer prior to completing a master's degree in education in 1998, at which time he became a schoolteacher. He was currently in his second marriage, which prior to his developing sexual preoccupations had been stable for 2 years.

> **BOX 5.4 CASE STUDY: PEDOPHILIA AND RIGHT ORBITOFRONTAL TUMOR (*Continued*)**
>
> During a neurologic examination, he solicited female team members for sexual favors. He was unconcerned that he had urinated on himself. He was slow to initiate leftward saccades and had mild left nasolabial fold flattening without facial weakness.
>
> Appendicular tone was increased bilaterally. There was no neglect. Abnormal glabellar, snout, and palmomental responses were present. The patient's gait was wide based, and as he walked, his step length diminished and side-to-side titubation occurred.
>
> Magnetic resonance imaging revealed an enhancing anterior fossa skull base mass that displaced the right orbitofrontal lobe.
>
>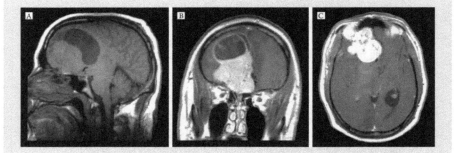
>
> **Photo 5.3** Magnetic resonance imaging scans at the time of initial neurologic evaluation: T1 sagittal (A), contrast-enhanced coronal (B), and contrast-enhanced axial (C) views. In A and B, the tumor mass extends superiorly from the olfactory groove, displacing the right orbitofrontal cortex and distorting the dorsolateral prefrontal cortex. (Image from: Burns, J. M., & Swerdlow, R. H. (2003). Right orbitofrontal tumor with pedophilia symptom and constructional apraxia sign. *Archives of Neurology*, 60(3), 437–440. doi:10.1001/archneur.60.3.437.)
>
> The patient's tumor was removed and, 7 months later, he was determined to no longer pose a threat to his stepdaughter and returned to his home. Approximately a year later, the patient again developed a persistent headache. He also began to secretly collect pornography. Magnetic resonance imaging showed tumor regrowth. The tumor was again resectioned.

Pedophiles

In contrast to the study on rapists, several studies suggest reduced activation in the subcortical and cortical regions in pedophiles on tasks of visual-erotic stimulation featuring adults.

Criminals in the Lab

An illustrative study conducted by Walter and colleagues matched 13 pedophiles with 14 controls. The pedophiles had each been convicted of crimes against at least one child under 10 years of age. Both groups were shown 256 images that included adult erotic images, emotionally laden images, and neutral images. Although the pedophiles demonstrated similar responses to controls for the nonerotic material, they exhibited significantly lower signal intensities when presented with the adult erotic stimuli. Specifically, the pedophile group demonstrated decreased activity in the hypothalamus, dorsolaterial, and dorsomedial prefrontal cortex regions, suggesting possible neural correlates to a lack of sexual interest toward adults.

White-Collar Criminals

The vast majority of laboratory research on criminal behavior has focused upon those who have engaged in violence. A 2012 study by Raine and colleagues, however, focused upon a destructive but typically non-physically violent criminal subtype: the white-collar criminal.

The white-collar criminal group was comprised of 21 individuals who had "used computers illegally to gain money or valuable information, cheated or conned a person, business, or government for gain; obtained unemployment or sickness benefit by telling lies, stolen supplies from work, using a stolen check or credit card, and lied about income on a tax return."

The subject group was matched with 21 criminal control subjects on basic demographics and general level of offending.

The researchers utilized both neuropsychological assessments and structural MRI. The latter revealed that the white-collar criminals had increased cortical thickness in several regions of the prefrontal cortex associated with planning, decision-making, conflict reasoning, social perception, and the ability to inhibit undesirable responses. Specifically, white-collar criminals showed increased cortical thickness in the left ventromedial prefrontal cortex, the right inferior frontal gyrus, the right precentral gyrus, the right postcentral, gyrus, the right posterior superior temporal gyrus, and the inferior parietal region of the right temporal–parietal junction.

Criminal Subtype Conclusions

In the studies cited, as well as others that have sought to illuminate the role of brain functioning in crime, the researchers note the limitations of neurocriminology: small sample sizes, the inability to apply general findings to specific individuals, the implication of the same regions of the brain in both criminal behavior and in noncriminal psychopathologies and compromised functioning.

Acknowledging the need for further research, as well as the aggravating or mitigating role of external factors such as genetics and environment, the laboratory studies on criminals have nonetheless revealed brain–criminal behavior correlates that have increased the body of knowledge surrounding factors that influence engagement in criminal behavior.

Key Terms

Continuous Performance Test (CPT): Any of several standardized tests that measure sustained attention, impulsivity, and dyscontrol. In normal controls, CPT engagement generally results in higher glucose activity in the frontal, temporal, and parietal lobes.

Predatory Violence: Instrumental violence, characterized by planning designed to accomplish a desired goal.

Affective Violence: Impulsive or reactive violence, characterized by unplanned and uncontrolled acts typically triggered by real or perceived insults or threats.

Parahippocampal Gyrus: Part of the limbic system that surrounds the hippocampus. Associated with emotional dysregulation, inappropriate responding to social cues, and impulsivity that could lead to violence.

Use Your Brain

Test Your Knowledge

1. Neuroscientific research on crime subtypes has typically involved:
   a. Imaging on a small subset of individuals
   b. Meta-analysis of studies related to the target behavior
   c. Case studies
   d. All of the above

2. In their research on murderers, Raine and colleagues found murderers to have abnormal asymmetries in glucose activity in the amydala, thalamus, and medial temporal gyrus. Specifically, these asymmetries involved:
   a. Relatively greater anterior activity and relatively reduced posterior activity
   b. Relatively greater posterior activity and relatively reduced anterior activity
   c. Relatively reduced left activity and relatively greater right activity
   d. Relatively reduced right activity and relatively greater left activity

Criminals in the Lab

3. Research has found that both affective and predatory murderers have significantly higher levels of right subcortical glucose metabolism. However:
   a. Affective murderers were found to have lower prefrontal activity, whereas predatory murderers were found to have prefrontal activity that was consistent with the non-murder control subjects.
   b. Predatory murderers were found to have lower prefrontal activity, whereas affective murderers were found to have prefrontal activity that was consistent with the non-murder control subjects.
   c. Affective murderers were found to have lower anterior activity, whereas predatory murderers were found to have posterior activity that was consistent with the non-murder control subjects.
   d. Affective murderers were found to have lower posterior activity, whereas predatory murderers were found to have anterior activity that was consistent with the non-murder control subjects.

4. Research on brain functioning related to rapists and pedophiles has found that:
   a. Pedophiles show decreased activity in the hypothalamus, dorsolaterial and dorsomedial prefrontal cortex regions, whereas rapists have heightened activity in these regions.
   b. Rapists show decreased activity in the hypothalamus, dorsolaterial and dorsomedial prefrontal cortex regions, whereas pedaphiles have heightened activity in these regions.
   c. No significant differences between pedophiles and rapists in relation to hypothalamus, dorsolaterial and dorsomedial prefrontal cortex region responses.
   d. Increased activity in relation to hypothalamus, dorsolaterial and dorsomedial prefrontal cortex region responses in both pedophiles and rapists.

5. As opposed to criminals who engage in physical violence, research has found white-collar criminals to have:
   a. Decreased cortical thickness in several regions of the prefrontal cortex.
   b. Decreased overall brain volume.
   c. Increased cortical thickness in several regions of the prefrontal cortex.
   d. Increased overall brain volume.

Answer Key:
1. (d) 2. (c) 3. (a) 4. (a) 5. (c)

Apply Your Knowledge

1. Research demonstrates differences in brain functioning between criminal subtypes. To what extent, if any, could this research be used to inform public policy related to criminal justice?
2. Several case studies have shown that it is possible for an individual without a prior criminal history to suffer brain dysfunction following tumor or injury, and subsequently engage in antisocial behavior, including crime. Does "acquired sociopathy" impact culpability? Why or why not?

Bibliography

Bannon, S., Salis, K., & O'Leary, K. D. (2015). Structural brain abnormalities in aggression and violent behavior. *Aggression and Violent Behavior, 25*, 323–331.

Blair, R. (2013). Psychopathy: Cognitive and neural dysfunction. *Dialogues in Clinical Neuroscience, 15*(2), 181–190.

Bueso-Izquierdo, N., Verdejo-Román, J., Contreras-Rodríguez, O., Carmona-Perera, M., Pérez García, M., & Hidalgo-Ruzzante, N. (2016). Are batterers different from other criminals? An fMRI study. *Social Cognitive and Affective Neuroscience, 11*(5), 852–862. doi:10.1093/scan/nsw020.

Burns, J. M., & Swerdlow, R. H. (2003). Right orbitofrontal tumor with pedophilia symptom and constructional apraxia sign. *Archives of Neurology, 60*(3), 437–440. doi:10.1001/archneur.60.3.437.

Chen, C., Raine, A., Kun-Hsien, C., Chen, I., Hung, D., & Ching-Po, L. (2016). Abnormal white matter integrity in rapists as indicated by diffusion tensor imaging. *BMC Neuroscience, 17*, 45. doi:10.1186/s12868-016-0278-3.

Cope, L., Ermer, E., Gaudet, L., Steele, V., Eckhardt, A., Arbabshirani, M., Caldwell, V., Calhoun, K., & Kiehl, K. (2014). Abnormal brain structure in youth who commit homicide. *NeuroImage: Clinical, 4*, 800–807.

Gregory, S., fftche, D., Simmons, A., Kumari, V., Howard, M., Hodgins, S., & Backwood, N. (2012). The antisocial brain: Psychopathy matters. *Archives of General Psychiatry, 69*(9), 962–972.

Joyal, C., Black, D., & Dassylva, B. (2007). The neuropsychology and neurology of sexual deviance: A review and pilot study. *Sexual Abuse, 19*, 155–173. doi:10.1007/s11194-0079045-4.

Lanata, S., & Miller, B. (2016). The behavioural variant frontotemporal dementia (bvFTD) syndrome in psychiatry. *Journal of Neurosurgical Psychiatry, 87*(5), 501–511. doi:10.1136/jnnp-2015-310697.

Molenberghs, P., Ogilvie, C., Louis, W. R., Decety, J., Bagnall, J., & Bain, P. G. (2015). The neural correlates of justified and unjustified killing: An fMRI study. *Social Cognitive and Affective Neuroscience, 10*(10), 1397–1404. doi:10.1093/scan/nsv027.

Paquola, C., Bennett, M., Hatton, S., Hermens, D., Groote, I., & Lagopoulos, J. (2017). Hippocampal development in youth with a history of childhood maltreatment. *Journal of Psychiatric Research, 9*, 149–155.

Criminals in the Lab

Raine, A., Buchsbaum, M., & Lacasse, L. (1997). Brain abnormalities in murderers indicated by positron emission tomography. *Biological Psychiatry, 42*, 495–508.

Raine, A., Meloy, J. R., Bihrle, S., Stoddard, J., LaCasse, L., & Buchsbaum, M. (1998). Reduced prefrontal and increased subcortical brain functioning assessed using positron emission tomography in predatory and affective murderers. *Behavioral Sciences and the Law, 16*, 319–332.

Raine, A., Laufer, W., Yang, Y., Narr, K., Thompson, P., & Toga, A. (2012). Increased executive functioning, attention, and cortical thickness in white-collar criminals. *Human Brain Mapping, 33*, 2932–2940.

Walter, M., Witzel, J., Wiebking, C., Gubka, U., Rotte, M., Schiltz, K., Bermpohl, F., Tempelmann, C., Bogerts, B., Heinze, H., & Northoff, G. (2007). Pedophilia is linked to reduced activation in hypothalamus and lateral prefrontal cortex during visual erotic stimulation. *Biological Psychiatry, 62*, 698–701.

Yang, Y., Raine, A., Chen-Bo, H., Schug, R., Toga, A., & Narr, K. (2010). Reduced hippocampal and parahippocampal volumes in murderers with schizophrenia. *Psychiatry Research: Neuroimaging, 182*, 9–13.

Neurocriminology in the Criminal Justice System

Prevention and Investigation

6

Organic matter, especially nervous tissue, seems endowed with a very extraordinary degree of plasticity.

William James, *The Principles of Psychology*

A liar should have a good memory.

Quintilian

Learning Objectives

1. Describe the influence of neuroscience on crime prevention.
2. Describe the current status of neuroscience efforts to support criminal investigations.
3. Analyze the potential and limitations of translating modern neuroscience to crime prevention and investigation efforts.

The application of neuroscience to outcome improvement has gained acceptance in numerous contexts including sports performance, treating psychosis, rehabilitating traumatic brain injury, and even advancing leadership development.

Applying neuroscience to the prevention and investigation of crime (and to prosecution and sentencing, as will be explored in Chapter 7) has met with greater controversy, reinvigorating metaphysical arguments regarding free will versus determinism, and associated sociopolitical debates regarding the appropriate balance between rehabilitation and retribution in criminal justice efforts. Despite neuroscientific advances unforeseeable at the time that Cesare Lombroso conducted his postmortem studies, the brain continues to resist providing the categorical answers—of guilty mind or not, will be violent or not, will recidivate or refrain from reoffense—that are at the center of most criminal justice proceedings. Yet, its more nuanced answers have not been without utility.

Although definitive answers to the complex questions of crime prevention, intervention, and control remain elusive, researchers engaged in translational neuroscience have suggested and demonstrated that current advances offer significant opportunities for the early identification, intervention, and potential prevention of crime.

Socially anchored criminology focused on managing criminogenic risk factors in the external environment—poverty, early exposure to violence, overcrowding—or upon clear and timely punishment to deter and control crime.

Neuroscience offers the possibility of the precise identification of the brain dysfunction associated with engagement in criminal activity, and the potential for a more targeted response.

The translation of neuroscience to the prevention and investigation phases of the criminal justice process is based on two contrasting hypotheses: (1) the brain's neuroplasticity can support behavioral change; and (2) the brain's stability can be excavated for objective truth.

The former hypothesis has proven particularly relevant to juvenile justice, early intervention programs, and leveraged therapeutic interventions, whereas the latter has been involved in efforts related to memory recovery and lie detection.

Prevention: The Brain Can Change

In 1989, two consolidated cases required the United States Supreme Court to decide if the capital punishment of a 16- or 17-year-old minor constituted cruel and unusual punishment under the Eighth Amendment of the U.S. Constitution.

In *Stanford v. Kennedy*, the late Justice Antonin Scalia, writing for the majority in the divided court decision, concluded that the petitioners' argument failed to establish that the execution of 16 and 17 year olds violates society's "evolving standards of decency."

Scalia added the following:

> We also reject petitioners' argument that we should invalidate capital punishment of 16- and 17-year-old offenders on the ground that it fails to serve the legitimate goals of penology. According to petitioners, it fails to deter because juveniles, possessing less developed cognitive skills than adults, are less likely to fear death; and it fails to exact just retribution because juveniles, being less mature and responsible, are also less morally blameworthy. In support of these claims, petitioners and their supporting *amici* marshal an array of socioscientific evidence concerning the psychological and emotional development of 16 and 17 year olds.
>
> If such evidence could conclusively establish the entire lack of deterrent effect and moral responsibility, resort to the Cruel and Unusual Punishments Clause would be unnecessary; the Equal Protection Clause of the Fourteenth Amendment would invalidate these laws for lack of rational basis.

Prevention and Investigation

BOX 6.1 STANFORD V. KENTUCKY

Prior to 2005, the Supreme Court had concluded that the execution of 16 and 17 year olds does not violate the Eighth Amendment, prohibition against cruel and unusual punishment. The decision was rendered in the consolidation of two cases, *Stanford v. Kentucky* and *Wilkins v. Missouri*.

STANFORD V. KENTUCKY

On January 7, 1981, 17-year-old Kevin Stanford and an accomplice raped, sodomized, and subsequently murdered a gas station attendant after committing a robbery that yielded approximately 300 cartons of cigarettes, 2 gallons of fuel, and a small amount of cash. A corrections officer later testified that Stanford said that he had to kill the victim because she lived next door to him and could identify him. Stanford had added, "... I guess we could have tied her up or something or beat [her up]... and tell her if she tells, we would kill her... Then after he said that he started laughing."

Based upon the seriousness of the offense, and a history of failed attempts by the Kentucky juvenile justice system to treat him for delinquency, Stanford was tried as an adult.

He was convicted of murder, first-degree sodomy, first-degree robbery, and receiving stolen property, and was sentenced to death and 45 years in prison. The Kentucky Supreme Court affirmed the death sentence, rejecting Stanford's "deman[d] that he has a constitutional right to treatment." Finding that the record clearly demonstrated that "there was no program or treatment appropriate for the appellant in the juvenile justice system," the court held that the juvenile court did not err in certifying the petitioner for trial as an adult. The court also stated that petitioner's "age and the possibility that he might be rehabilitated were mitigating factors appropriately left to the consideration of the jury that tried him."

WILKINS V. MISSOURI

Four years later, 16-year-old Heath Wilkins and an accomplice robbed a convenience store, stabbing the 26-year-old mother of two who worked the counter. When the accomplice had trouble opening the register and the woman spoke up to assist, Wilkens stabbed her three more times in the heart, and four times in the neck, opening her carotid artery. The robbery yielded liquor, cigarettes, rolling papers, and approximately $450. Records indicate that Wilkins had planned

(Continued)

114 Neurocriminology

> ### BOX 6.1 STANFORD V. KENTUCKY (*Continued*)
>
> to rob the store and murder whoever was behind the counter because "a dead person can't talk."
>
> As the Kentucky juvenile court had done with Stanford, the Missouri juvenile system certified Wilkins as an adult based upon the viciousness of the crime and the failure of the juvenile system to rehabilitate him following previous delinquent acts.
>
> Wilkins was convicted of murder, first-degree sodomy, first-degree robbery, and receiving stolen property. He pleaded guilty to first-degree murder, armed criminal acts, and carrying a conceal weapon.
>
> Both Stanford and Wilkins were sentenced to death. Each appealed, and the Supreme Court granted certiorari.
>
> In their appeal the petitioners argued that capital punishment of 16 and 17 year olds is contrary to "evolving standards of decency that mark the progress of a maturing society."
>
> The Court disagreed. Writing for the majority, Justice Antonin Scalia stated: "Of the 37 States whose laws permit capital punishment, 15 decline to impose it upon 16-year-old offenders and 12 decline to impose it on 17-year-old offenders. This does not establish the degree of national consensus this Court has previously thought sufficient to label a particular punishment cruel and unusual."
>
> Sixteen years later, national consensus and that of the Court shifted, attributed in part to neuroscientific advances related to adolescent brain functioning.

Sixteen years later, the Court would reach a different conclusion. Included in the amicus testimony considered was research that utilized fMRI scans to assess the differences between adolescent and adult brain functioning.

Roper v. Simmons

At the age of 17, Christopher Simmons enlisted two friends—aged 15 and 16—to commit murder. Simmons told the two that he wanted to break and enter a victim's home, commit burglary, tie the victim up, and throw the victim from a bridge. He stated that the three could "get away with it" because they were minors.

Prior to the commission of the crime, the 16-year-old friend left; conspiracy charges would later be brought against him, then dropped in exchange for his testimony against Simmons.

As planned, on September 8, 1993 at approximately 2:00 am, Simmons and the 15 year old entered the home of the female victim, who Simmons

Prevention and Investigation

recognized from a car accident that had involved them both. Simmons would later confess that his recognition of the victim confirmed his resolve to complete the murder. The two teens used duct tape to cover the victim's eyes and mouth and to bind her hands before taking her in her minivan to a state park. They covered her head with a towel, walked her to a railroad trestle bridge that spanned a river and, after tying her hands and feet together with electrical wire, threw her from the bridge, drowning her.

On the afternoon of September 9th, fishermen recovered the victim's body.

In the interim, Simmons bragged about the killing to his friends. On September 10th, police arrested him at his school. Simmons waived his right to an attorney; after less than 2 hours of interrogation, Simmons confessed to the murder.

Simmons was charged with burglary, kidnapping, stealing, and murder in the first degree. He was tried as an adult. The jury found him guilty and the State sought the death penalty.

For a defendant to be eligible for the death penalty, the judge or jury must find one aggravating factor and consider all relevant mitigating evidence. Aggravating factors can be in the definition of the crime (such as murder of a police officer or murder of a child), or in separate sentencing factors (such as the offender's probability of posing a continued risk of violence), or both. Mitigating evidence considers broad factors related to an offender's history, current status, and probable future violence risk.

As aggravating factors, the State submitted that the murder was committed for money (the burglary); was committed for the purpose of avoiding, interfering with, or preventing lawful arrest of the defendant (murdering the victim to prevent her identification him); and involved depravity of mind and was outrageously and wantonly vile, horrible, and inhuman.

As mitigating evidence, Simmons's attorneys offered Simmons's lack of prior criminal history and close family and community ties. During closing arguments, the defense also offered Simmons's age as a mitigating factor: "the legislatures have wisely decided that individuals of a certain age aren't responsible enough" to drink, serve on juries, or even see certain movies and, therefore, age should make "a huge difference to [the jurors] in deciding just exactly what sort of punishment to make."

In rebuttal, the prosecutor gave the following response: "Age, he says. Think about age. Seventeen years old. Isn't that scary? Doesn't that scare you? Mitigating? Quite the contrary I submit. Quite the contrary."

The jury recommended the death penalty, a recommendation that the trial judge accepted.

Simmons sought to have his conviction and sentence set aside based upon ineffective assistance at trial. His contention included testimony by evaluating clinical psychologists who concluded that Simmons was "very

116 Neurocriminology

immature," "very impulsive," and "very susceptible to being manipulated or influenced." The experts testified that Simmons's background included a difficult home environment, and dramatic behavioral changes that began during adolescence and included poor school performance, long absences from home, and alcohol and drug use with other teenagers and young adults. Simmons's new defense cited each of these as mitigating factors that should have been established in the sentencing proceeding.

The trial court denied the defense's motion, a denial affirmed by the Missouri Supreme Court. The federal courts also denied Simmons's petition for a writ of habeas corpus in 2001.

Then, in 2002, the U.S. Supreme Court decided *Atkins v. Virginia*, and held that the Eighth and Fourteenth Amendments prohibit the execution of individuals who suffer from what was then diagnosed as mental retardation (now known as intellectual developmental disorder). Of significance to the Simmons case, the Court based its decision in part on a determination that those so diagnosed "frequently know the difference between right and wrong and are competent to stand trial, but, by definition, they have diminished capacities to understand and process information, to communicate, to abstract from mistakes and learn from experience, to engage in logical reasoning, to control impulses, and to understand others' reactions. Their deficiencies do not warrant an exemption from criminal sanctions, but diminish their personal culpability."

Simmons filed a new petition for state postconviction relief, arguing that the reasoning applied in the *Atkins* case should extend to those under age 18. The Missouri Supreme Court agreed and set aside Simmons death sentence, resentencing him to life without possibility of parole. The State appealed to the U.S. Supreme Court, which granted certiorari in 2004.

The American Psychological Association and the Missouri Psychological Association jointly filed one of the amicus curiae in support of abolishing the death penalty for those under 18 years of age. Included in the brief were arguments supported by neuroscientific research:

> Recent research suggests a biological dimension to adolescent behavioral immaturity: the human brain does not settle into its mature, adult form until after the adolescent years have passed and a person has entered young adulthood. Advances in magnetic resonance imaging (MRI) technology have opened a new window into the differences between adolescent and adult brains. MRI technology produces exquisitely accurate pictures of the inner body and brain. Beginning in the 1990s, "functional" MRIs have allowed mapping not only of brain anatomy but observation of brain functioning while an individual performs tasks involving speech, perception, reasoning, and decision-making...Of particular interest with regard to decision-making and criminal culpability is the development of the frontal lobes of the brain. The frontal lobes, especially the prefrontal cortex, play a critical role in the executive or

Prevention and Investigation

"CEO" functions of the brain which are considered the higher 10 functions of the brain... They are involved when an individual plans and implements goal-directed behaviors by selecting, coordinating, and applying the cognitive skills necessary to accomplish the goal... Emerging from the neuropsychological research is a striking view of the brain and its gradual maturation, in far greater detail than seen before. Although the precise underlying mechanisms continue to be explored, what is certain is that, in late adolescence, important aspects of brain maturation remain incomplete, particularly those involving the brain's executive functions.

Although the Court did not specifically reference the neuroscientific testimony, the Simmons case has been cited as a milestone for the introduction of neuroscientific evidence in the courts. Writing for the majority, Justice Anthony Kennedy stated: "as any parent knows and as the scientific and sociological studies respondent and his *amici* cite tend to confirm, '[a] lack of maturity and an underdeveloped sense of responsibility are found in youth more often than in adults and are more understandable among the young. These qualities often result in impetuous and ill-considered actions and decisions'. Once the diminished culpability of juveniles is recognized, it is evident that the penological justifications for the death penalty apply to them with lesser force than to adults."

Kennedy added,

Although the Court cannot deny or overlook the brutal crimes too many juvenile offenders have committed, it disagrees with petitioner's contention that, given the Court's own insistence on individualized consideration in capital sentencing, it is arbitrary and unnecessary to adopt a categorical rule barring imposition of the death penalty on an offender under 18...When a juvenile commits a heinous crime, the State can exact forfeiture of some of the most basic liberties, but the State cannot extinguish his life and his potential to attain a mature understanding of his own humanity. While drawing the line at 18 is subject to the objections always raised against categorical rules, that is the point where society draws the line for many purposes between childhood and adulthood and the age at which the line for death eligibility ought to rest.

Five years later, Justice Kennedy would write a majority opinion that would again link youth, neuroscience, and the potential for cognitive and affective development.

Graham v. Florida

Terrance Graham was born to crack-addicted parents whose drug use persisted into his early years. He was diagnosed with attention-deficit hyperactivity disorder (ADHD) in elementary school, began drinking alcohol and using tobacco at age 9, and smoked marijuana at age 13. In July 2003, at age 16, Graham and three peers attempted to rob a restaurant at which one

118   Neurocriminology

of the peers worked. The employed peer left a back door unlocked, through which Graham and his accomplice entered. The two wore masks. The accomplice struck the manager in the back of the head with a metal bar. When the manager yelled, the accomplice and Graham ran to the waiting car of a third peer. No money was taken from the restaurant.

Graham was arrested and tried as an adult for armed burglary with assault or battery, which carried a maximum penalty of life without the possibility of parole, and attempted armed-robbery, which carried a maximum penalty of 15 years imprisonment. Graham pleaded guilty to both charges under an agreement that included a letter to the court. Graham wrote, "I made a promise to God and myself that if I get a second chance, I'm going to do whatever it takes to get to the [National Football League]."

The trial court accepted the plea, and withheld adjudication of guilt on both charges. Graham was sentenced to concurrent 3-year terms of probation.

Less than 6 months later, at age 17, Graham was again arrested. His was charged with participating in a home invasion robbery, possessing a firearm, and associating with persons engaged in criminal activity, which were clear violations of his probation.

Under Florida law, the recommended minimum sentence Graham could receive was 5 years' imprisonment. The maximum was life. Graham's attorney requested the minimum of 5 years. A presentence report prepared by the Florida Department of Corrections recommended a downward departure of 4 years' imprisonment. The State recommended that Graham receive 30 years on the armed burglary count plus 15 years on the attempted armed-robbery count.

The presiding judge, who was different than the one who had heard Graham's first case, offered an explanation for the sentence he would pronounce:

> Mr. Graham, as I look back on your case, yours is really candidly a sad situation. You had, as far as I can tell, you have quite a family structure. You had a lot of people who wanted to try and help you get your life turned around including the court system, and you had a judge who took the step to try and give you direction through his probation order to give you a chance to get back onto track. And at the time you seemed through your letters that that is exactly what you wanted to do. And I don't know why it is that you threw your life away. I don't know why…Given your escalating pattern of criminal conduct, it is apparent to the Court that you have decided that this is the way you are going to live your life and that the only thing I can do now is to try and protect the community from your actions.

The court found Graham guilty of his earlier armed burglary and attempted armed robbery charges, and sentenced him to life and 15 years, respectively. Under Florida law, Graham did not have the possibility of parole absent executive clemency. He filed a motion to challenge his sentence under the

Prevention and Investigation 119

Eighth Amendment, which was denied by the trial court. The denial was affirmed by the District Court, which concluded that Graham's sentence was not grossly disproportionate to his crimes and that he was incapable of rehabilitation.

The Supreme Court disagreed. Justice Kennedy stated,

> The inadequacy of penological theory to justify life without parole sentences for juvenile nonhomicide offenders, the limited culpability of such offenders, and the severity of these sentences all lead the Court to conclude that the sentencing practice at issue is cruel and unusual. No recent data provide reason to reconsider *Roper*'s holding that because juveniles have lessened culpability they are less deserving of the most serious forms of punishment.

Chief Justice Roberts, in a concurring opinion, added the following:

> As petitioner's *amici* point out, developments in psychology and brain science continue to show fundamental differences between juvenile and adult minds. For example, parts of the brain involved in behavior control continue to mature through late adolescence. Juveniles are more capable of change than are adults, and their actions are less likely to be evidence of "irretrievably depraved character" than are the actions of adults. It remains true that "[f]rom a moral standpoint it would be misguided to equate the failings of a minor with those of an adult, for a greater possibility exists that a minor's character deficiencies will be reformed."

The court reversed and remanded the trial court's decision.

In 2012, the Court further extended its evolving position on the culpability of minors to the adjudication of minors who commit homicide.

Miller v. Alabama

In *Miller v. Alabama*, the Supreme Court heard two consolidated cases of 14 year olds convicted of homicide and sentenced to life in prison without the possibility of parole.

The first case involved Kuntrell Jackson who, in November 1999, robbed a video store with two peers. En route to the store, Jackson learned that one of the boys carried a sawed-off shotgun in his coat sleeve. Jackson decided to stay outside when the two other boys entered the store. Inside, the boy with the shotgun pointed the gun at the store clerk, demanding money. She refused. A few moments later, Jackson entered the store to find the peer with the gun continuing to demand money. At trial, the parties disputed whether he warned the clerk that "[w]e ain't playin'," or instead told his friends, "I thought you all was playin'." When the clerk threatened to call the police, the peer shot and killed her.

Jackson was charged with capital felony murder and aggravated robbery. He was tried as an adult, convicted, and sentenced to life without parole.

The second case involved Evan Miller. Miller's mother suffered from alcoholism and drug addiction and his stepfather abused him. He had been in and out of foster care during his 14 years. He regularly used drugs and alcohol and he had attempted suicide four times, with the first attempt at age six. In 2003, Miller was at home with a friend when a neighbor arrived to make a drug deal with Miller's mother. The two boys followed the neighbor back to his trailer, where the three smoked marijuana and drank. When the neighbor passed out, Miller stole his wallet, splitting about $300 with his friend. Miller tried to put the wallet back in the neighbor's pocket. The neighbor awoke and grabbed Miller by the throat. Miller's friend hit the neighbor with a baseball bat. Once released, Miller grabbed the bat and repeatedly struck the neighbor with it before placing a sheet over the neighbor's head, telling him, "I am God, I've come to take your life," and delivered an additional blow. The boys then returned to Miller's home, before deciding to return to the neighbor's trailer to conceal their crime. Once there, they lit two fires; the neighbor died from his injuries and smoke inhalation.

Miller was initially charged as a juvenile, and then transferred to adult court where he was charged with murder in the course of arson. Miller was found guilty and, like Jackson, sentenced to life without the possibility of parole.

In writing the majority opinion in which the Court held that mandatory sentences life without parole for those under the age of 18 is unconstitutional, Justice Elena Kagan applied reasoning that was similar to that employed in the prior decisions:

> Our decisions rested not only on common sense—on what "any parent knows"—but on science and social science as well. In *Roper*, we cited studies showing that "'[o]nly a relatively small proportion of adolescents'" who engage in illegal activity "'develop entrenched patterns of problem behavior.'" ... And in *Graham*, we noted that "developments in psychology and brain science continue to show fundamental differences between juvenile and adult minds"—for example, in "parts of the brain involved in behavior control." We reasoned that those findings—of transient rashness, proclivity for risk, and inability to assess consequences—both lessened a child's "moral culpability" and enhanced the prospect that, as the years go by and neurological development occurs, his "'deficiencies will be reformed.'"

Through these rulings, the Court accepted the neuroscientific conclusion that the adolescent brain undergoes changes that have significant ramifications for redemption and rehabilitation.

Concurrent advances in neuroscience explored its potential to move beyond providing explanations for brain–behavior maturation to informing interventions that prevent initial criminal engagement and recidivism in youth.

Prevention and Intervention

The Supreme Court's rulings related to the sentencing of youth convicted of violent offenses are partially based on research that demonstrates that executive functioning is not fully developed until the mid-twenties.

The rulings are also based upon contemporary research related to the brain's neuroplasticity—the ability to form and reorganize neural connections in response to learning or experience (see Box 6.2). Neuroplasticity is at the nexus of neurocriminology's contributions to prevention and intervention efforts in the criminal justice arena, particularly in relation to leveraged treatment typically specified by problem-solving courts.

BOX 6.2   NEUROPLASTICITY

Brain plasticity, or neuroplasticity, refers to the brain's ability to change and adapt.

It is a process that starts at birth.

Infant neurons have an estimated 2,500 synapses, which increase through synaptogenesis to approximately 15,000 by age 3. The increase coincides with infants' independent exploration of the world and associated learning of new skills, which in turn increase the brain's structure, functions, and connections. Through a process of synaptic pruning, the number of synapses halves by adulthood, as more functional—or used—connections are strengthened and less functional ones atrophy, thereby increasing neural efficiency.

Neuroplasticity is most prominent when the brain is developing, and is influenced by a variety of factors, including genetics, injury, and disease, as well as external factors, including learning opportunities, exaggerated stress exposure, malnutrition, and substance abuse.

External influences have been shown to have particular ramifications for both the development of antisocial behavior and rehabilitation; neuroscience has focused upon identifying the specific factors associated with the development of antisocial traits and on interventions that can modify neural circuits to strengthen brain functions that support the regulation and control of antisocial behavior. These have been incorporated into evidence-based interventions that support modern prevention and intervention efforts focused upon deterring juvenile crime.

Although strongest during the developmental period, the brain's neuroplasticity has been proven to remain responsive to experience and learning in at least certain regions of the brain throughout the life span.

(Continued)

BOX 6.2 NEUROPLASTICITY (*Continued*)

In a 2006 study, for example, researchers Eleanor Maguire, Katherine Woollett, and Hugo Spiers found that the licensed London taxi drivers, when compared with a similarly matched group of London bus drivers, showed greater gray matter volume in the posterior hippocampi; the flexible navigation skills required by the taxi drivers correlated with increases in the brain region associated with learning and using spatial representation.

Research of that same year conducted by Bogdon Draganski and colleagues found that gray matter volume increased in the posterior hippocampus of medical students as they studied for their medical exam. Most significantly, the increases did not abate 3 months following the examination, suggesting that the learning was associated with structural gray matter changes.

These and other findings underlie neuroscientific efforts to assist in "rewiring" neural responses that are associated with substance abuse and aggression, offering potential additional intervention options to the specialty criminal justice courts that focus upon the rehabilitation of targeted criminal subtypes to support the prevention of recidivism.

The Brain and Early-Onset Antisocial Behaviors

The Court's decision to distinguish between youth and adult culpability in relation to criminal behavior and, consequently, to limit maximum sentences that youth can receive for specific crimes reference modern neuroscience as an informing factor. The "youth-sensitive" approach, however, dates to 1899 when the first U.S. juvenile court determined that a newly named group of individuals—adolescents—lacked the full maturity, reasoning ability, and impulse control to be fully responsible for their actions. Retribution was designed to be closely paired with rehabilitation that included mandated social activities, insight-oriented or behavioral-based psychotherapeutic interventions, and eventually diversion that precluded association with the criminal justice system altogether. Research on the impact on the juvenile justice system was often compromised by the variability of the programs offered, as well as methodological flaws in research designs. The ambiguity of results precipitated a reconsideration of the separate system for juvenile offenders. Calls for revisiting the more lenient approach to juvenile justice were amplified in the 1980s, which saw a rise in both juvenile and adult crime rates, characterizations of some urban youth as "superpredators," and petitions for "adult time for adult crime."

Prevention and Investigation

The influence of these sociopolitical factors on juvenile justice was significant: Rates of juvenile detention increased by 600% between 1977 and 1986.

By the mid-1990s, new research emerged that pointed to the adverse impacts of these efforts. Youth placed in adult facilities were significantly more likely to be victims of abuse, violence, and rape. Additionally, and contrary to aims to improve public safety, studies found that adolescents who were transferred to the adult system were more likely to recidivate than adolescents charged with the same crimes who remained under juvenile court jurisdictions.

The neuroscience that, in part, informed the Supreme Court's sentencing decisions related to adolescent offenders is also providing research that may support prevention and early intervention of youth violence.

For the vast majority of youth, early engagement in risky behaviors is the result of asymmetries in brain development: The development of the limbic system typically outpaces that of the prefrontal cortex (PFC). As a result, early adolescents tend to be emotionally driven; they seek sensation and gratification and lack the parallel ability to control their impulses or think through the possible consequences of their actions. Over the course of the teen years, the normally developing brain will experience an increase in the inhibitory connections between these two regions, which gradually assist youth to regulate impulsivity and affect, and consider the social consequences of their actions.

Compromises in normal brain development, whether resulting from genetic, neurobiological, or environmental factors, can delay or halt this important developmental process.

Evidence of such compromises often pre-dates adolescence.

Research has demonstrated that delays in executive functioning in early years tend to grow larger as children age. Children who show an early inability to regulate their emotions or behavior, or to engage in good decision-making, or to conform to social norms, are generally unable to acquire these abilities independently, leading to greater risk of engagement in antisocial activity including crime.

The correlation between frontal lobe dysfunction and antisocial behavior is one of the best replicated brain-imaging findings of modern neuroscience. Studies have consistently found that behavioral and emotional regulation depend upon effective communication between the orbitofrontal cortex (OFC), the anterior cingulate cortex (ACC), the dorsolateral prefrontal cortex (DLPFC), and the amygdala. As noted in previous chapters, the OFC is involved in emotional processing, learning from reward and punishment, and decision-making. The ACC is involved in conflict management and error processing. The DLPFC has been implicated behavioral regulation and attentional and cognitive flexibility. The amygdala is involved in emotional responsiveness and fear conditioning.

Prevention and Early Intervention Efforts

Preliminary efforts to directly employ neuroscience to inform the prevention of criminal behavior in youth have primarily centered upon these brain regions to support improvement of regulatory control.

fMRI studies, for example, have measured the impact of attention training on youth diagnosed with ADHD, which is characterized by impulsivity, inattention, and hyperactivity, and has high comorbidity rates with conduct and oppositional defiant disorders (characterized by patterns of violating age-appropriate norms and angry/irritable mood or argumentative/defiant behavior, respectively). Computer-based, game-like activities designed to improve short-term memory, visual processing, and eye-hand coordination have been correlated with decreased ADHD symptomology.

An increasing body of research has focused upon mindfulness training, such as training in meditation, as a path to improving cognitive performance and decision-making, emotion regulation, and stress resilience among youth. Recent studies have suggested that mindfulness training may strengthen youth's ability to volitionally switch attention, which could potentially provide much needed time to consider more adaptive solutions when confronting adversity. Favorable results have also been found with neurofeedback, which measures brain waves in real-time, allowing for the provision of visual and auditory feedback to help develop personal control over focus.

Structural MRI studies in youth with conduct disorder have shown significant correlations between decreased gray matter volume in the PFC and the amygdala, and the severity of conduct problems. Additionally, brain-imaging research has suggested a positive correlation between omega-3 fatty-acid supplementation and increased PFC functioning. Based upon these findings, researcher Adrian Raine and colleagues conducted a study of 200 children between the ages of 8 and 16 to determine if simply providing supplementation could be impactful in improving youth behavioral issues. Parents who participated in the study gave their children a daily drink for a 6-month period. Half of the participants ingested a drink that was fortified with 1000 mg of omega 3.

Although reports of child externalizing and internalizing aggressive and antisocial behaviors were insignificant between the two groups at the conclusion of the 6-month trial, longer term results (6 months following trial conclusion) showed that children who received the supplementation demonstrated reduced parent-rated aggression. Significantly, the research also found reductions in psychopathy and reactive aggression among the parents of children who ingested the supplements.

These findings are consistent with those of other researchers, such as Rachel Gow and colleagues, who found an inverse relationship between

Prevention and Investigation

callous–unemotional and antisocial traits—associated with psychopathy—and omega-3 levels.

Drug Courts

According to the Bureau of Justice Statistics Report, two-thirds of the 2.3 million inmates incarcerated in the United States met the clinical criteria for addiction or substance abuse in the year prior to their arrest. Nearly 18% committed crimes to procure drugs. Nearly one-third (32%) of state prisoners and more than one-quarter (26%) of federal prisoners reported that they had committed their index offense while under the influence of drugs. Among state prisoners, drug offenders (44%) and property offenders (39%) reported the highest incidence of drug use at the time of offense. Among federal prisoners, drug offenders (32%) and violent offenders (24%) reported the highest incidences.

An estimated 95% of offenders resume drug abuse following release from prison. An estimated 60%–80% recidivate.

Given the strong correlation between drug abuse and crime, improving treatment outcomes is particularly significant to public health and safety.

An important contribution to this effort occurred in 1989 when Florida established the nation's first drug court. In the intervening years, the National Institute of Justice (NIH) estimates that more than 3,000 Drug Courts have been established in the United States. Although target populations, structures, and resources vary, these courts generally involve a standardized assessment of an offender's risk, needs, and potential responsivity to treatment, coupled with judicial interaction, monitoring, sanctions, incentives, and treatment and rehabilitation services.

A 5-year longitudinal study, which sampled nearly 1,800 drug and nondrug court probationers, found that drug courts were correlated with significant reductions in relapse and criminal behavior. Specifically, the average recidivism rate for drug court participants was 16% in the first year after leaving the program, and 27% after the second year, compared to 46% and 60%, respectively, for nondrug court participants. Interestingly, results held across multiple populations, including those with antisocial personality disorder, though they were reduced for those with comorbid narcissism or depression.

Despite their effectiveness, fewer than 10% of substance-abusing criminals are adjudicated through drug courts, suggesting the need for additional access, as well as additional evidence-based alternatives.

The Brain and Drug Abuse

Decades of research has supported the influence of genetics on substance abuse; studies of twins, siblings, and other family members have found that genetics, at the population level, account for up to 50% of vulnerability, or brain bias, for substance use and abuse.

Neuroimaging studies have advanced knowledge regarding the neurobehavioral bases of substance abuse, that is, of the specific regions of the brain that may be implicated in drug abuse vulnerabilities. As with antisocial behavior, substance abuse is associated with dysfunctional interactions between the PFC and subcortical areas of the brain, including the ACC. In substance abusers, hyperactivity in the cingulate cortex, associated with reward seeking and craving, appear to diminish the cognitive and behavioral regulatory inputs of the PFC. Compared to nondrug users, those who abuse drugs show hyperactivity during drug craving and hypoactivity during withdrawal, suggesting a stronger experience of need and a greater intolerance for abstinence.

Additional studies have implicated the dopaminergic system—the brain's natural reward circuitry—and have suggested that those who abuse drugs may initially do so to increase low dopamine levels. Commonly abused drugs trigger significant neural activity that flood the accumbens nucleus of the limbic system by as much as ten times that experienced by naturally occurring pleasurable experiences, creating a powerful motivation to continue use. Over time, the effected brain adapts to the higher levels of dopamine and competing natural pleasures become even less salient. This adaptation also triggers aversive physical withdrawal symptoms when the drug is not ingested. The powerful motivation for reward, coupled with the aversive withdrawal symptoms, contribute to the difficulty many addicts experience in breaking the cycle of addiction.

Preliminary Findings

Given the similarities in the brain regions associated with substance abuse and antisocial behavior, it is unsurprising that neuroscientifically informed prevention and early intervention efforts have adopted similar protocols.

For example, mindfulness-based approaches have been successfully applied to support those addicted to cocaine. Cocaine abusers participated in a PET scan study during which they were instructed to cognitively inhibit craving while watching videos of individuals preparing drug paraphernalia and smoking crack cocaine. The abusers showed lower activity in the right OFC and right accumbens when engaged in efforts to inhibit cravings compared with activity levels on scans taken when not engaged in such attempts, suggesting the ability to volitionally control cravings.

Real-time fMRI, or rt-fMRI, is another method by which researchers are exploring ways to improve outcomes for those who experience drug addiction. As with neurofeedback techniques, rt-MRI provides subjects with a contemporaneous analysis of their brain data to support self-regulation of brain activations in response to drug cues—basically, with the ability to use feedback to retrain one's brain response to cravings.

Prevention and Investigation

BOX 6.3   THE DRUG BRAIN ON TRIAL

Prior to and since the establishment of drug specialty courts, the justice system has commonly used sobriety as a condition of parole and probation. Those who violate this condition risk re-incarceration.

Neuroscientific research has confirmed that specific regions of the brain are associated with and compromised by repeated use and abuse of substances, and that the powerful impact on the dopaminergic system in particular contributes to the difficulty many addicts experience when seeking to break the cycle of addiction. Recent research, including studies involving the use of real-time MRIs, has demonstrated that those with substance addiction can volitionally control cravings and may be able to "retrain" the brain to support sobriety.

Both aspects of these neuroscientific findings were introduced in contrasting amici curiae submitted in a highly controversial case heard by the Massachusetts Supreme Judicial Court in late 2017.

In *Commonwealth v. Eldred,* the court was asked to consider whether imposing a drug free condition on probationers diagnosed with substance abuse disorder violates these individuals' Eighth Amendment protection against cruel and unusual punishment.

The question stems from a case involving defendant Julie Eldred, who was charged with stealing jewelry from a dog-walking client to fund her drug use. Her case was continued without a finding, and she was placed on probation with a condition that she remains drug-free and consent to random drug screenings.

Two weeks after her sentencing, Eldred tested positive for Fentanyl, an opioid that was originally manufactured for use as an anesthetic during surgery. Its only acceptable prescription use is for break-through cancer pain, but several investigations have concluded that it is frequently prescribed "off-label." It is also sold on the black market in pure form or laced with heroin.

Eldred was found to be in violation of the terms of her probation and ordered held in jail until she could be placed in an in-patient treatment program.

Eldred's attorney argued that ordering Eldred to remain drug-free was akin to ordering her to put a "chronic brain disease" in remission.

Her position was supported in the amicus curiae submitted by the Massachusetts Medical Society, which was joined by 34 other interests including the American Academy of Addiction Psychiatry and Northeastern University School of Law's Center for Health Policy and Law.

(*Continued*)

BOX 6.3   THE DRUG BRAIN ON TRIAL (*Continued*)

The brief incorporates neuroscientific evidence to support the conceptualization of substance use disorder (SUD) as a brain disease, particularly given the alterations that substances cause to neural circuitry. The brief also argues that, as a result of these changes, relapse is a typical attribute of SUD and, as such, should not be considered merely a failure of will. Citing research, the authors further suggest that requiring abstinence as a condition of parole or probation may undermine the very outcome sought as the threat of reincarceration creates a contraindicated stress condition:

> [E]motional stressors... trigger heightened activity in brain stress circuits. The anatomy (the brain circuitry involved) and the physiology (the neurotransmitters involved) have been delineated through neuroscience research. Relapse triggered by exposure to stressful experiences involves brain stress circuits beyond the hypothalamic-pituitary-adrenal axis that is well known as the core of the endocrine stress system. There are two of these relapse-triggering brain stress circuits- one originates... in the brainstem... the other originates in the central nucleus of the amygdala.

The brief concludes by urging the court to "take into account the scientific consensus that SUD is a chronic disease of the brain" and that relapse is a "symptom" for which reincarceration is an inappropriate response.

In contrast, a second amici curiae submitted by 11 addiction experts including Gene Heyman of Boston College, Scott O. Lilienfeld of Emory University, Stephen Morse of the University of Pennsylvania, and Sally Satel of Yale University suggests a different interpretation of the scientific evidence.

The brief includes cautions related to limitations commonly associated with the current state of neuroscientific evidence generally: Research typically involves small samples of non-randomly selected individuals. Lack of replication studies has been conducted to demonstrate reliability. The correlation between brain activity and behavior does not demonstrate causation, which is particularly salient in relation to those with SUD given the prevalence of comorbidity with other mental health disorders that could conflate findings.

The authors also cite the socio–cultural–political influence on efforts to characterize substance abuse disorder: "Efforts to position addiction as a 'brain disease' were intended to persuade politicians

(Continued)

Prevention and Investigation 129

BOX 6.3 THE DRUG BRAIN ON TRIAL (*Continued*)

and society to take the problem seriously other than as a moral failure. The model's appeal is obvious: It is tidy. It signifies medical gravitas and neuroscientific sophistication."

The brief continues:

> In the hands of those who subscribe to and promote the brain-disease model, brain imaging is often intended as a visual refutation of the existence of the addict's capacity to refrain from using substances. In a typical imaging experiment conducted with positron emission tomography (PET) or functional magnetic resonance imagining (fMRI), addicts watch videos of people handling a crack pipe or needle, causing their prefrontal cortices, amygdala, and other brain structures to activate beyond the base rate of activity in the region of interest (the entire brain is active all the time) (Goldstein & Volkow, 2002). Videos of neutral content, such as landscapes, induce no such heightened response while the brains of comparison subjects presented these stimuli are being scanned. The resultant Technicolor images of affected brain regions, which are simply graphic representations of complex mathematical data and are not "pictures" of the brain, are undeniably arresting. These images are meant to convince us that the mere will to change or choice in the face of rewards or punishment cannot be expected to override these tissue or physiological changes. After all, it appears that one can "see" the damage inflicted on the now allegedly "broken" brain.

The authors suggest that neuroscience, in fact, supports the continuation of sobriety as a condition of parole and probation:

> The mere association of drug taking with expected neurobiological changes in the brain is not evidence that drug use is beyond control. This is abundantly evident from the large volume of data demonstrating that addiction is a set of behaviors whose course can be altered by foreseeable consequences. The same cannot be said of conventional brain diseases such as Alzheimer's or multiple sclerosis. In sum, the best scientific and clinical data are strongly at odds with the view that addicts are unable to choose not to use substances. We believe that a decision in favor of the probationer could have significant, even devastating, implications for the future of treatment-based approaches to criminal justice as well as for criminal responsibility more generally. We conclude that the probationer's claim should be denied because it is based on erroneous, refuted scientific premises and will have negative consequences if it is accepted.

(Continued)

> **BOX 6.3   THE DRUG BRAIN ON TRIAL (*Continued*)**
>
> The Eldred case is significant for its consideration of neuroscientific evidence to illuminate dynamics important to the issue at hand. In relation to substance use, which has widespread public health and safety ramifications, neuroscience has confirmed the powerful impact of substances upon the brain, as well as the possibility and challenges associated with breaking the cycle of addiction.
>
> The Eldred case also suggests that this evidence can inform, but not answer, ultimate legal questions, such as whether requiring substance abusers to remain substance-free violates their constitutional rights.

Veterans Treatment Courts

In 2008, the first Veterans Treatment Court (VTC) was established in Buffalo, New York. VTCs are a combination of specialty drug and mental health courts that target former military service members. In addition to drug and other supervised treatment, these programs engage an interdisciplinary team that includes the judge, volunteer veteran mentors, family members, and community service organizations. Additionally, the team includes the prosecutor and defense attorney, who suspend traditional adversarial roles to support defendant success. The defendant is closely monitored and supervised by the court, and subject to random drug and alcohol testing. As of 2017, more than 300 VTCs were established nationwide.

A 2016 report by the National Drug Court Institution (NDCI) found that the limited available research on VTCs suggests that they prevent recidivism and support adaptive functioning through improved substance use, psychiatric symptoms, and social and family relationships.

In addition to recognizing the role of traumatic brain injury in altering behavior, VTCs often consider the impact of post-traumatic stress disorder (PTSD) on functioning, a mental illness that has not generally mitigated culpability during the guilt phase except in extreme cases of domestic violence.

The Diagnostic and Statistical Manual of Mental Disorders-5 (DSM-5) characterizes PTSD as a trauma and stressor-related disorder that follows exposure to actual or threatened death, serious injury, or sexual violence. Symptoms include recurrent, involuntary, and intrusive distressing memories; recurrent distressing dreams, dissociative reactions such as flashbacks; and/or distress at internal or external triggers. Symptoms can also include avoidance behaviors associated with the traumatic event and negative alterations in cognition, mood, and/or arousal responses.

Prevention and Investigation

Studies estimate that national prevalence rates for violence among individuals with PTSD are 7.5%. In contrast, prevalence rates for violence among post-9/11 veterans with PTSD are estimated at 19.5%. These elevated prevalence rates coincide with relatively disproportionate rates of incarceration; although 7.3% of Americans have served in the military at some point in their lives, veterans represent an estimated 10% of inmates.

Two primary reasons have been associated with the willingness of the justice system—and the public—to accept PTSD and a mitigating factor in determining the criminal responsibility of veterans: (1) claims of PTSD are perceived to be more credible when made by those who have served in the military, and (2) given veteran engagement in criminality is influenced—if not directly caused by—service to the government, it is palatable that the government is permitted to treat rather than punish those affected by this service. Several states have codified this sentiment in legislation that allows the court to divert veterans to treatment rather than incarceration if PTSD arising from military service is demonstrated to be a contributing factor to their crime.

Neuroscience has sought to advance the understanding of PTSD through the identification of associated structural abnormalities and functional changes in brain regions. Studies commonly implicate the PFC, hippocampus, and amygdala. A 2011 meta-analysis conducted by researchers Katherine Hughes and Lisa Shin, for example, found that individuals with PTSD generally demonstrate hyperresponsivity in the amygdala, which is correlated with an exaggerated fear response. In contrast, PTSD is associated with decreased activity in the medial PFC, which compromises emotional and behavioral regulation and decision-making. The hippocampus has also been implicated, resulting in impairments in the consolidation and retrieval of trauma memories, which may be associated with an inability to process, contextualize, and manage traumatic experiences. Research results related to the nature of hippocampal abnormality vary, with some suggesting increased activity and others suggesting decreased depending upon the experimental methodology employed.

Several challenges have been associated with translating these findings into intervention strategies similar to those suggested for juvenile and drug specialty courts. Principally, the current limited understanding the potential interplay between TBI and PTSD, which are often comorbid in post–9/11 veterans, renders it difficult to theorize a generalizable treatment strategy. This challenge notwithstanding, it is likely that—consistent with efforts related to other specialty courts—the growing body of research related to brain functioning and veterans will be introduced to inform and influence VTCs.

BOX 6.4   LEGISLATING LENIENCY

Public and legal acceptance of the correlation between post–9/11 military service, PTSD, and criminality have led several states to legislate options for the adjunction of veterans who engage in criminal activity.

California and Minnesota were the first states to codify the link between PTSD and veteran PTSD.

CALIFORNIA PENAL CODE § 1170.9

(a) In the case of any person convicted of a criminal offense who could otherwise be sentenced to county jail or state prison and who alleges that he or she committed the offense as a result of sexual trauma, traumatic brain injury, PTSD, substance abuse, or mental health problems stemming from service in the United States military, the court shall, prior to sentencing, make a determination as to whether the defendant was, or currently is, a member of the United States military and whether the defendant may be suffering from sexual trauma, traumatic brain injury, PTSD, substance abuse, or mental health problems as a result of his or her service. The court may request, through existing resources, an assessment to aid in that determination.

(b) (1) If the court concludes that a defendant convicted of a criminal offense is a person described in subdivision (a), and if the defendant is otherwise eligible for probation, the court shall consider the circumstances described in subdivision (a) as a factor in favor of granting probation.

(2) If the court places the defendant on probation, the court may order the defendant into a local, state, federal, or private nonprofit treatment program for a period not to exceed that period which the defendant would have served in state prison or county jail, provided the defendant agrees to participate in the program and the court determines that an appropriate treatment program exists.

(c) If a referral is made to the county mental health authority, the county shall be obligated to provide mental health treatment services only to the extent that resources are available for that purpose, as described in paragraph (5) of subdivision (b) of Section 5600.3 of the Welfare and Institutions Code. If mental health treatment services are ordered by the court, the county mental health agency shall coordinate appropriate referral of the defendant to the county veterans service officer, as described in paragraph (5) of subdivision (b) of Section 5600.3 of the Welfare and Institutions Code. The county mental health agency

(Continued)

Prevention and Investigation 133

BOX 6.4 LEGISLATING LENIENCY (*Continued*)

shall not be responsible for providing services outside its traditional scope of services. An order shall be made referring a defendant to a county mental health agency only if that agency has agreed to accept responsibility for the treatment of the defendant.

(d) When determining the "needs of the defendant," for purposes of Section 1202.7, the court shall consider the fact that the defendant is a person described in subdivision (a) in assessing whether the defendant should be placed on probation and ordered into a federal or community-based treatment service program with a demonstrated history of specializing in the treatment of mental health problems, including substance abuse, PTSD, traumatic brain injury, military sexual trauma, and other related mental health problems.

(e) A defendant granted probation under this section and committed to a residential treatment program shall earn sentence credits for the actual time the defendant serves in residential treatment.

(f) The court, in making an order under this section to commit a defendant to an established treatment program, shall give preference to a treatment program that has a history of successfully treating veterans who suffer from sexual trauma, traumatic brain injury, PTSD, substance abuse, or mental health problems as a result of that service, including, but not limited to, programs operated by the United States Department of Defense or the United States Department of Veterans Affairs.

(g) The court and the assigned treatment program may collaborate with the Department of Veterans Affairs and the United States Department of Veterans Affairs to maximize benefits and services provided to the veteran.

(h) (1) It is in the interests of justice to restore a defendant who acquired a criminal record due to a mental health disorder stemming from service in the United States military to the community of law abiding citizens. The restorative provisions of this subdivision shall apply to cases in which a trial court or a court monitoring the defendant's performance of probation pursuant to this section finds at a public hearing, held after not less than 15 days' notice to the prosecution, the defense, and any victim of the offense, that all of the following describe the defendant:

(A) He or she was granted probation and was at the time that probation was granted a person described in subdivision (a).

(B) He or she is in substantial compliance with the conditions of that probation.

(Continued)

BOX 6.4   LEGISLATING LENIENCY (*Continued*)

(C) He or she has successfully participated in court-ordered treatment and services to address the sexual trauma, traumatic brain injury, PTSD, substance abuse, or mental health problems stemming from military service.

(D) He or she does not represent a danger to the health and safety of others.

(E) He or she has demonstrated significant benefit from court-ordered education, treatment, or rehabilitation to clearly show that granting restorative relief pursuant to this subdivision would be in the interests of justice.

(2) When determining whether granting restorative relief pursuant to this subdivision is in the interests of justice, the court may consider, among other factors, all of the following:

(A) The defendant's completion and degree of participation in education, treatment, and rehabilitation as ordered by the court.

(B) The defendant's progress in formal education.

(C) The defendant's development of career potential.

(D) The defendant's leadership and personal responsibility efforts.

(E) The defendant's contribution of service in support of the community.

(3) If the court finds that a case satisfies each of the requirements described in paragraph (1), then the court may take any of the following actions by a written order setting forth the reasons for so doing:

(A) Deem all conditions of probation to be satisfied, including fines, fees, assessment, and programs, and terminate probation prior to the expiration of the term of probation. This subparagraph does not apply to any court-ordered victim restitution.

(B) Reduce an eligible felony to a misdemeanor pursuant to subdivision (b) of Section 17.

(C) Grant relief in accordance with Section 1203.4.

(4) Notwithstanding anything to the contrary in Section 1203.4, a dismissal of the action pursuant to this subdivision has the following effect:

(A) Except as otherwise provided in this paragraph, a dismissal of the action pursuant to this subdivision releases the defendant from all penalties and disabilities resulting from the offense of which the defendant has been convicted in the dismissed action.

(B) A dismissal pursuant to this subdivision does not apply to any of the following:

(*Continued*)

Prevention and Investigation 135

BOX 6.4   LEGISLATING LENIENCY (*Continued*)

(i) A conviction pursuant to subdivision (c) of Section 42002.1 of the Vehicle Code.

(ii) A felony conviction pursuant to subdivision (d) of Section 261.5.

(iii) A conviction pursuant to subdivision (c) of Section 286.

(iv) A conviction pursuant to Section 288.

(v) A conviction pursuant to subdivision (c) of Section 288a.

(vi) A conviction pursuant to Section 288.5.

(vii) A conviction pursuant to subdivision (j) of Section 289.

(viii) The requirement to register pursuant to Section 290.

(C) The defendant is not obligated to disclose the arrest on the dismissed action, the dismissed action, or the conviction that was set aside when information concerning prior arrests or convictions is requested to be given under oath, affirmation, or otherwise. The defendant may indicate that he or she has not been arrested when his or her only arrest concerns the dismissed action, except when the defendant is required to disclose the arrest, the conviction that was set aside, and the dismissed action in response to any direct question contained in any questionnaire or application for any law enforcement position.

(D) A dismissal pursuant to this subdivision may, in the discretion of the court, order the sealing of police records of the arrest and court records of the dismissed action, thereafter viewable by the public only in accordance with a court order.

(E) The dismissal of the action pursuant to this subdivision shall be a bar to any future action based on the conduct charged in the dismissed action.

(F) In any subsequent prosecution for any other offense, a conviction that was set aside in the dismissed action may be pleaded and proved as a prior conviction and shall have the same effect as if the dismissal pursuant to this subdivision had not been granted.

(G) A conviction that was set aside in the dismissed action may be considered a conviction for the purpose of administratively revoking or suspending or otherwise limiting the defendant's driving privilege on the ground of two or more convictions.

(H) The defendant's DNA sample and profile in the DNA data bank shall not be removed by a dismissal pursuant to this subdivision.

(I) Dismissal of an accusation, information, or conviction pursuant to this section does not authorize a defendant to own, possess, or have in his or her custody or control any firearm or prevent his or her

(Continued)

> ### BOX 6.4   LEGISLATING LENIENCY (*Continued*)
>
> conviction pursuant to Chapter 2 (commencing with Section 29800) of Division 9 of Title 4 of Part 6.
>
> #### MINNESOTA 609.115 PRESENTENCE INVESTIGATION § SUBD. 10. MILITARY VETERANS
>
> (a) When a defendant appears in court and is convicted of a crime, the court shall inquire whether the defendant is currently serving in or is a veteran of the armed forces of the United States.
>
> (b) If the defendant is currently serving in the military or is a veteran and has been diagnosed as having a mental illness by a qualified psychiatrist or clinical psychologist or physician, the court may:
>
> (1) order that the officer preparing the report under subdivision 1 consult with the United States Department of Veterans Affairs, Minnesota Department of Veterans Affairs, or another agency or person with suitable knowledge or experience, for the purpose of providing the court with information regarding treatment options available to the defendant, including federal, state, and local programming; and
>
> (2) consider the treatment recommendations of any diagnosing or treating mental health professionals together with the treatment options available to the defendant in imposing sentence.

Neurocriminology and Specialty Courts: Possibilities and Limitations

Neuroscience represents an additional tool that can inform prevention and treatment of select populations—most specifically juveniles, those addicted to illicit substances, and potentially veterans—by offering targeted interventions and an additional means to measure outcomes. Theoretically, one can conduct a pre- and post-neuroimage to determine the impact of an intervention on brain functioning. If the dysfunction correlated with behavior is valid and reliable, and the brain dysfunction is mitigated post-intervention, the behavior will presumably be favorably altered. Given the high economic cost of incarceration, and the high economic and emotional costs of recidivism, such efforts represent a worthwhile strategy.

The limitations of current neuroscientific research in informing criminal justice responses generally—small sample sizes, general-to-individual translation limitations—are one impediment to this potential being realized at the present time.

Prevention and Investigation 137

These limitations are compounded when addressing juvenile criminality by the very factor that renders intervention potentially more favorable: The naturally occurring changes in brain functionality during adolescence add to the heterogeneity of factors that could contribute to an individual youth's engagement in criminal behavior.

For example, as in adults, a distinction can be made between adolescent aggression that is affective/impulsive versus purposeful/instrumental. In affective/impulsive aggression, a youth responds to a stimulus—being called a name, being bumped into—that is perceived as threatening or provocative. In youth, the response could result from hyperactivity in the amygdala or delayed functioning in the PFC, which would warrant different treatment interventions. Purposeful/instrumental aggression—punching a peer to steal a coveted object—may demonstrate of amygdala hypofunction, resulting in problems in using emotional responsiveness to learn new behavior, or decreased activity in the OFC, which compromises the ability to achieve goals through socially acceptable means. Determining the implicated brain region likewise has important ramifications for developing the most efficacious intervention.

Additionally, the high rates of comorbidity with youth antisocial behavior and other conditions, ranging from oppositional defiant disorder and ADHDs, to mood disorders such as anxiety and depression, further suggests a need to identify reliable brain–behavior subtypes associated with juvenile crime to support prevention and intervention efficacy.

Comorbidity is likewise a challenge for adults, and the same regions implicated in addiction and PTSD are also implicated in other disorders, such as depression and anxiety.

Measuring individual outcomes, such as is done in medical settings that employ neuroimaging, may also be cost prohibitive, particularly without widespread public support for a criminal justice system that generally favors treatment over retribution.

These limitations notwithstanding, the developmental trajectory and neuroplasticity of the brain are and will likely remain foundational to efforts to inform and support crime prevention.

In contrast, a presumption of constancy is at the core of the translation of neuroscience to crime investigation.

Neuroscience and Criminal Investigation

Science and the criminal justice system seek truth. Although approaches and priorities are different, both have focused considerable effort on two areas of shared interest: memory and deception.

Science has historically offered three main techniques to assist the criminal justice system in discerning truth: verbal response analysis, such

138   Neurocriminology

as Criteria-Based Content Analysis (CBCA) and reality monitoring (RM) techniques; paraverbal response analysis, including micro-facial expression analysis; and physiological response measurement, most routinely measured via the polygraph (see Box 6.5).

BOX 6.5   PSYCHOLOGICAL SCIENCE AND LIE DETECTION

Psychological science has offered three main techniques for discriminating between truthful and deceptive statements: psychophysiological response measurement, verbal response analysis, and paraverbal response analysis. While none of these techniques is foolproof, each offers the trained practitioner the potential to increase his or her discernment ability beyond the 50% accuracy that would occur by chance.

The most common psychophysiological response measurement tool is the polygraph, a combination of medical devices that measure an individual's heart rate, blood pressure, respiratory rate, and electrodermal activity (such as sweat), and monitor changes from an established baseline. In North America, the typical polygraph examination includes a pretest interview during which the examiner or forensic psychophysiologist explains the polygraph procedure and gains preliminary information that will be used as the basis for the Control Question Test. During the testing phase, the examiner asks control questions (e.g., "Is your name John Smith?" "Are you 45 years of age?") to which most individuals will respond honestly. The responses to these questions are used to establish an individual baseline. Fluctuations from the baseline when questions related to a crime or incident are posed ("Did you murder the victim?" "Did you hide evidence of your crime?") reflect a stress reaction that may signify deception. During the third, post-test phase of the examination, the examiner interprets the results to determine if the interviewee has been deceptive.

Several factors compromise the reliability of the polygraph. Stress responses, for example, can result from a variety of emotions; a fluctuation may signify anger, anxiety, or fear rather than deception, rendering a false positive result (i.e., the response of a truthful person is determined to be deceptive). Poorly crafted control questions can also influence reliability. For example, questions such as "Have you ever been in trouble with the law?" or "Have you ever hit your daughter?" can result in false positives when dealing with an individual who has a stress response to the question content, or false negatives in individuals who do not consider traffic violations unlawful or spanking a form of hitting. Suspects can also attempt countermeasures, such as the use of sedatives

(Continued)

BOX 6.5 PSYCHOLOGICAL SCIENCE AND LIE DETECTION (*Continued*)

or the application of antiperspirant to the fingertips to relax or interrupt a sweat response. Suspects have also placed sharp objects in a shoe or bitten their tongues, lips, or cheeks in an attempt to distract from stressful questions or to ensure that reactions to all questions evoke identical physiological responses. Research also suggests that certain test subjects, such as psychopaths, do not exhibit an arousal response to stress and would therefore result in false negative responses to the examination.

According to the National Research Council, the polygraph "can discriminate lying from truth telling at rates well above chance, though well below perfection." Across studies, the polygraph has been found to have between an 80% and 90% accuracy rate when administered by the skilled examiner. The subjectivity involved in interpreting exam results, coupled with the less than 90% accuracy rate, renders the polygraph generally inadmissible in court (save in cases in which both parties agree or the presiding judge allows admission).

Notable techniques that improve discernment of falsehood via verbal communication include the CBCA and the RM technique.

CBCA was originally developed to evaluate children's statements in sexual abuse cases but has also been utilized as a tool to identify deception by adults. CBCA is based on the premise that only a person who has actually experienced an event will be able to produce a statement with the characteristics that are described in the CBCA criteria. Specifically, these criteria include general characteristics (i.e., the narrative of an actual event will be logical and coherent, the interviewee will digress or shift focus at points, and the narrative will be significantly detailed). Truthful statements must also meet a substantial number of specific content criteria (contextual embedding, interactions, reproduction of speech, unexpected complications, unusual details, superfluous details, accurately reported details misunderstood, related external associations, subjective experience, attribution of the accused's mental state) and motivation-related criteria (spontaneous corrections or additions, admitting lack of memory or knowledge, raising doubts about one's own testimony, self-deprecation, and pardoning the accused). Accuracy rates of discerning truth from falsehood when utilizing CBCA have ranged from 65% to 80%.

A second notable technique used in verbal analysis of lie detection is the RM technique. RM is based on the premise that memories based on experienced events differ in quality from memories of fabricated events. Real experiences are perceived through the five senses of sight,

BOX 6.5 PSYCHOLOGICAL SCIENCE AND LIE DETECTION (*Continued*)

hearing, smell, taste, and touch. Memories of real experiences, therefore, are more likely to contain perceptual information, such as details related to smell ("He had a very noticeable body odor"), taste ("He gagged me with a rag that tasted like cologne"), touch ("The ground beneath me was rough and uneven"), as well as specific visual and auditory details ("There was light from a nearby streetlamp" and "I could hear a toilet flush somewhere in the house"). Experienced memories also tend to include contextual information, such as spatial and temporal details ("He stood right next to me, leaning into my shoulder" and "He stopped the car, then yanked the door open, then grabbed my arm"). Accounts of imagined events, in contrast, are the result of internal reasoning and are more likely to include cognitive statements such as "He must have had a knife because I was stabbed" or "I know he took my purse because it was gone when he left." Consistent with accuracy rates of CBCA, research has found RM to be accurate between 65% and 80% of the time.

There is no documented facial expression that correlates with guilt. In the analysis of nonverbal or paraverbal communications, however, investigators observe a person's behavior to infer whether he or she is lying. The analysis of nonverbal communication is based on the premise that because lying is cognitively more complex than telling the truth, those who are being deceptive will have uncontrolled behaviors as they focus on fabricating their narrative. In other words, the liar is so busy constructing his lie that he doesn't have the resources to control his mannerisms. Researcher Paul Ekman indicates that behavioral cues indicative of deception include higher, faster, and louder speech; greater pupil dilation; and fewer hand movements to accompany and illustrate speech. Liars who have not carefully prepared their narratives, and need to think carefully about their lies as they tell them, may speak more slowly than truth tellers.

The ability to detect deception through nonverbal communication is also influenced by the relationship between the liar and the interviewer, the interviewer's degree of familiarity with the liar's normal behavior, the interviewer's familiarity with the situation, the number of times the liar is interviewed, the liar's motivation to lie, and the interviewer's expectations.

Although these techniques improve the ability to discern accuracy above the level of chance, none offers a definitive way of determining whether a suspect or witness if providing reliable testimony—a possibility now being explored by neuroscience.

Prevention and Investigation 141

Although each of these methods can assist the trained practitioner, none reliably improves the ability to detect deception beyond the 50% accuracy that would occur by chance. Consequently, these techniques are rarely admissible in court under the *Frye* or *Daubert* standards of evidence, which require general acceptance by the scientific community, or general acceptance in addition to being valid, reliable, and having been subject to professional peer review, respectively.

BOX 6.6   CRITERIA FOR ADMISSIBLE EVIDENCE

Two cases generally guide state and federal decision-making in relation to the admission of evidence.

States that apply the "Frye Standard" are guided by the decision in *Frye v. United States,* 293 F. 1013 (D.C. Cir. 1923). In this case, a defendant convicted of second-degree murder appealed on the basis that the trial court did not admit scientific testimony related to a systolic blood pressure deception test—a predecessor to the polygraph—that the defendant underwent prior to trial and that he claimed proved that he was telling the truth regarding his innocence. In rejecting his appeal, the Court stated the following:

> Just when a scientific principle or discovery crosses the line between the experimental and demonstrable stages is difficult to define. Somewhere in this twilight zone the evidential force of the principle must be recognized, and while courts will go a long way in admitting expert testimony deduced from a well-recognized scientific principle or discovery, the thing from which the deduction is made must be sufficiently established to have gained general acceptance in the particular field in which it belongs.

Although some states continue to apply the general acceptance standard, many states and the federal courts have adopted the Daubert standard, which stems from the 1992 case, *Daubert v. Merrell Dow Pharmaceuticals, Inc.* (509 U.S. 579, 113 S.Ct. 2786). In this case, the petitioners were parents of two children born with birth defects that the parents believed resulted from first trimester ingestion of the antinausea drug Bendectin. To support their position, the petitioners retained experts who produced laboratory and reanalyzed epidemiological studies related to the drug. In contrast, Dow's expert reviewed all existing studies and research on human trials and concluded no association between drug ingestion and birth defects. Based upon

(Continued)

BOX 6.6 CRITERIA FOR ADMISSIBLE EVIDENCE (*Continued*)

their expert's findings, Dow petitioned for and was granted a summary judgment (a ruling that there are no factual issues to be tried and, therefore, the complaint can be decided by the court without trial).

Upon appeal, the Supreme Court remanded, and in doing so further informed guidelines for the admissibility of scientific evidence, concluding:

> "General acceptance" is not a necessary precondition to the admissibility of scientific evidence under the Federal Rules of Evidence, but the Rules of Evidence—especially Rule 702—do assign to the trial judge the task of ensuring that an expert's testimony both rests on a reliable foundation and is relevant to the task at hand. Pertinent evidence based on scientifically valid principles will satisfy those demands.

The five factors considered by the Daubert standard are as follows:

1. Whether a method can or has been tested
2. The known or potential rate of error
3. Whether the methods have been subjected to peer review
4. Whether there are standards controlling the technique's operation
5. The general acceptance of the method within the relevant community

The search for truth in the criminal justice system often includes reliance upon the testimony of suspects, accomplices, victims, and witnesses, which is directly tested by cross-examination. This approach can establish credibility or unearth inconsistencies, establishing if the testimony is convincing. It cannot, however, reveal if the testimony is based on an actual memory, or if it is an account designed to deceive.

The limitations of both science and the criminal justice system in consistently obtaining accurate and truthful testimony is evident in the significant role that false confessions and inaccurate witness testimony play in wrongful convictions that are subsequently overturned—estimated at 70% by the Innocence Project.

Given the critical importance of testimony to criminal justice proceedings, translating neuroscience to memory recovery and lie detection have been understandable priorities for research and investment.

BOX 6.7 THE WITNESS PROBLEM

The Innocence Project, founded in 1992 by Peter Neufeld and Barry Scheck at the Cardozo School of Law, has shown that eyewitness misidentification—that is, falsely remembered individuals or events—is the greatest contributing factor to convictions later proven wrong by DNA testing.

Numerous scientific studies have expounded upon the myriad of "memory myths." Some of the most relevant to the criminal justice system include:

Confidence Equals Accuracy: A "credible" witness is often defined by the faith he or she exhibits in the reported memory. Despite this, research (e.g., Kassin et al.) consistently demonstrates a lack of correlation between the confidence of the witness and the accuracy of the memory.

False Memories Are Uncommon: Although maintaining a complete fabrication to be true is rare in a cooperative witness, vulnerability to memory distortion—including significant memory distortion—is quite common. Studies have shown that memory can be influenced by the way in which a question is phrased. Subjects in a mock jury study, for example, remembered a significant number of unstated items regarding a robbery one week after hearing the testimony, demonstrating the extent to which we use preexisting "templates" to encode information, which may distort memory for new material (Holst et al.).

The Way a Question Is Posed Only Influences Vulnerable Witnesses: The use of non-neutral words, such as "claims" instead of "stated" and "denies" instead of "said" influence the way in which objective observers process information. Indeed, research has found that even nonverbal communication offered through body language or facial expressions can reinforce or compromise witness recall (Semmler et al.).

Racial Bias Is Only Relevant for the Racially Biased: Inaccuracy when remembering the faces of those of race or ethnic backgrounds different from one's own is well documented, regardless of bias or prejudice (Meissner et al.).

Since its founding, the Innocence Project has facilitated over 350 DNA exonerations nationwide. Eyewitness misidentification played a role in more than 70% of the convictions that were overturned.

This is one indication of the potential relevance of current neuroscientific research on memory, and the possible applications to the criminal justice system.

144   Neurocriminology

Witness Memory Recovery

More than any other brain function, memory is associated with our experience of individuality and uniqueness. It is multidimensional and represents the particular interdependence of our perceptions, experience, expectations, and learning.

The case of H.M. (highlighted in Chapter 2), studies of others who have suffered lesions and injuries, and more recent functional magnetic resonance imaging studies have revealed the ways in which memories are formed and retrieved.

When declarative memories—explicit memories for facts and events that can be consciously recalled and that are most relevant to testimony—are first created, or encoded, the medial temporal lobe (hippocampus and the entorhinal, perirhinal, and parahippocampal cortices) and relevant neocortical sensory regions (e.g., the visual cortex if sight is involved, the auditory cortex if sound is involved, and the amygdala if emotion is involved) are activated. Once a memory is fully formed, or consolidated, the medial temporal lobe is no longer active; the neocortex independently activates during remembering, along with the neocortical regions that initially processed and stored what was learned.

Retrieving a memory activates neural pathways that are distinguishable from the neural pathways that are activated when processing new information.

The development of a valid and reliable procedure for determining when an individual is remembering material versus encountering it for the first time would be of significant benefit during criminal investigations, aiding both suspect interrogations and witness interviews; the veracity of assertions as to whether people, items, or events were previously encountered would be "neurally confirmed."

Several studies have advanced this possibility.

In one published in 2010, researchers Jesse Rissman, Henry Greely, and Anthony Wagner showed 16 subjects a large set of face images. The subjects underwent fMRI scans while reviewing 400 face images that included face images from the original set as well as new face images. The subjects were asked to rate their recognition of the faces along of continuum that ranged from highly confident that they recalled the image to highly confident that the face image was new to them. Applying a statistical technique known as multivoxel pattern analysis (MVPA) to the data, trained classifiers were able to detect discernible differences in the neural activity associated with subjective reports of recalled versus new faces with between 70% and 90% accuracy.

The researchers next conducted an experiment to determine if such patterns could be discerned independent of subjective reports. To do so,

Prevention and Investigation

seven participants initially reviewed a set a face images before undergoing fMRI scans during which they were to indicate the gender of the image; recognition was not explicitly probed, but rather implicitly assessed. Additional scans replicated the explicit—or cued—recall procedure of the first experiment.

The results for the explicit recall were consistent with those in the initial experiment; differences in the neural activity associated with subjective reports of recalled versus new faces could be distinguished. However, the classifiers could not apply the same analysis to determine if subjects were viewing a familiar or new face image when the subjects themselves were not explicitly engaged memory retrieval: If the subject didn't indicate whether he remembered the face, the classifier couldn't determine if he did.

The findings show promise for the utilization of MVPA to fMRI data to verify the subjective report of eyewitness testimony, which suggests utility with cooperative witnesses. Conversely, the procedure was not able to reliably reveal whether an individual had experienced a particular stimulus—person, event, object—independent of a subjective report, suggesting a significant limitation when working with the often uncooperative participants in criminal justice proceedings.

Five years following their original study, Wagner and Rismman joined researchers Melina Uncapher, J. Tyler Boyd-Meredith, and Tiffany Chow and demonstrated the extent to which an uncooperative subject could feign or conceal memory.

Over the course of 2 days, 24 subjects participated in the new study. During day 1, the participants underwent fMRI scanning while engaging in an explicit memory task consistent with the one conducted in the original experiment; the subjects memorized 200 face images and then indicated those they remembered versus those that were new when shown a series of 400 face images. Also consistent with the initial research, the fMRI data were measured utilizing MVPA, which a trained classifier was able to decode with a mean accuracy of 63% across trials, which rose to 73% when the classifier reported the greatest confidence.

During day 2 of the experiment, the subjects were instructed to purposefully "trick" the computer, while appearing to cooperate. Specifically, the subjects were told to engage in cognitive tasks designed to conceal their memories of known face images and feign memories for new ones. When confronted with a known image, the subjects were instructed to immediately neutralize their memory by focusing upon elements of the image to which they had not previously attended, such as lighting, exposure, or other photographic or technical properties. In contrast, when confronted with a new image, they were instructed to immediately associate the image with someone they knew, and relive memories associated with that individual.

146   Neurocriminology

The results of the experiment unearthed a current limitation of applying fMRI scanning to witness memory detection: As opposed to the accuracy with which classifiers were able to decode subjective reports of actual memories, rates of accuracy for decoding concealed or feigned memories were at or less than chance.

The study showed that individuals can willfully activate or deactivate the regions of the brain associated with memory, and "beat the scan"—at least at present.

Given the vital importance of obtaining accurate information to successful investigations and just convictions, it is inevitable that research in this area will continue, as it has in the related domain of detecting deception.

Lie Detection

In 2002, two studies conducted in different parts of the world suggested that the process of deception might, indeed, be associated with an identifiable neural response.

In one of the studies, researchers at the University of Pennsylvania employed the Guilty Knowledge Test. The Guilty Knowledge Test measures psychophysiological responses to multiple choice questions that include options that are relevant to an event—such as an aspect of an investigation—and control alternatives that are unrelated to the event. Typically, individuals with foreknowledge—"guilty knowledge"—of the event will exhibit greater physiological responses to the relevant alternative than to the control alternatives.

Participants were instructed to memorize a specific playing card and told that they would receive a reward if they could successfully conceal their knowledge of the card when a series of cards were displayed on a computer. The participants underwent fMRI scans. Analyzing the resulting BOLD contrasts, researchers found that increased activity in the ACC, the superior frontal gyrus, and the left parietal cortex was associated with deceptive responses. They concluded that deception is detectable by fMRI.

The second experiment was conducted at the University of Hong Kong. In this experiment, participants were asked to feign memory problems and deliberately do poorly on a forced-choice test in which they were asked to remember one of two stimuli that had been previously presented to them. Consistent with findings employing the Guilty Knowledge Test, fMRI results showed evidence of increased prefrontal, subcortical, and parietal activity when deception was attempted.

Later studies attempted to improve the ecological validity of these research designs. A 2006 study, for example, randomly assigned subjects to one of two groups, a "guilty group" and a "non-guilty group." The guilty group was told that they would engage in a mock shooting, during which they would fire a

Prevention and Investigation

gun inside a hospital. The only one who knew of their involvement was the individual who provided the gun. The goal for each participant in the guilty group was to fool researchers into believing that he or she did not fire the weapon, even though the researchers suspected guilt based upon video surveillance that placed the participant at the scene at the time of the shooting. Video surveillance likewise placed the participants of the non-guilty group at the scene. However, the goal for each participant in this group was to be cooperative and truthful, and to demonstrate to the researchers that they did not, in fact, fire the gun.

An interesting element of the research design included having participants engage in a simulated shooting by firing blanks from a firearm prior to their interviews, rendering the guilty responses more akin to "true lies" that activate sensory and memory brain functioning. fMRI analysis revealed that deception was associated with increased activity in 14 areas in the frontal (left, medial, left inferior, bilateral central gyri); temporal (right hippocampus, right middle temporal gyrus); parietal (bilateral precuneus, right inferior parietal lobule); and occipital (left lingual gyrus) lobes, as well as in the anterior and posterior cingulate, the right fuisform gyrus, and the right sublobar insula.

These findings are consistent with meta-analyses on the use of fMRI to determine the correlation between brain functioning and deception: It appears that PFC—specifically the dorsolateral PFC, bilateral regions of the ventrolateral PFC, anterior insula, and the right ACC—are activated during attempts to lie (Photos 6.1 and 6.2).

The engagement of these regions comports with the theoretical construct of deception: To engage in deception requires working memory (associated with dorsolateral PFC) to be mindful of the truth while formulating a lie; the demonstration of inhibitory control to suppress a truthful response (associated with the ventrolateral PFC); and the regulation of affect while mentally task switching from the truth to the lie (implicating the anterior insula, and the right ACC).

The acceptance of fMRI lie detection in the courts has been mixed. In 2010, two unsuccessful attempts were made.

In *Wilson v. Corestaff Servs*, plaintiff Wilson sought an evidentiary hearing on the admissibility of fMRI to bolster the credibility of a key witness in a retaliation lawsuit. Wilson had been placed in a temporary work assignment at a bank and had complained of sexual harassment after an employee faxed an offensive nude photo to her workstation. She reported the incident to both the bank and Corestaff Services, her temporary employment agency. A principle at the agency allegedly told a witness not to place Wilson at further work assignments following her complaint. As the sole witness to the retaliation, the witness's credibility was a critical issue.

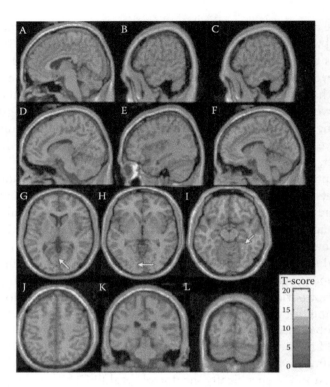

Photo 6.1 Functional MR images for lie condition in mock shooting study, which show activation in, A, anterior cingulate (sagittal section); B, left inferior frontal gyrus (sagittal section); C, left precentral gyrus (sagittal section); D, precuneus (sagittal section); E, inferior parietal lobule (sagittal section); F, sublobar insula or thalamus (sagittal section); G, posterior cingulate (transverse section [arrow]); H, left lingual gyrus (transverse section [arrow]); I, right fusiform gyrus (transverse section [arrow]); J, left medial frontal gyrus (transverse section); K, right hippocampus (coronal section); and L, right middle temporal (transverse section).

In applying the Frye standard used by New York State, the court rejected the fMRI evidence on two bases: (1) That the credibility of a fact witness is solely within the purview of the jury, and (2) That "the scientific literature raises serious issues about the lack of acceptance of the fMRI test in the scientific community to show a person's past mental state or to gauge credibility" and "fMRI test is akin to a polygraph test which has been widely rejected by New York State courts."

The second test of the court admissibility of fMRI for lie detection applied the Daubert standard. In *U.S. v Semrau*, the government charged psychologist Lorne Semrau with money laundering and health care fraud stemming from alleged purposeful "upcoding" of billing provided by the psychiatrists he employed to provide medication management and

Prevention and Investigation

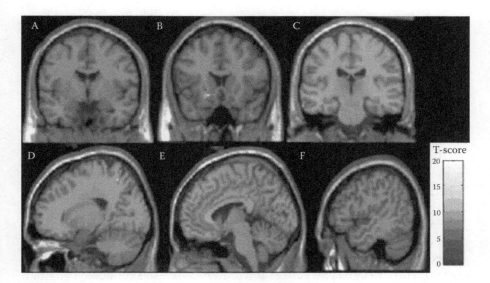

Photo 6.2 Functional MR images for the truth condition in the mock shooting study which show activation in *A*, precentral gyrus (transverse section); *B*, subcallosal gyrus or lentiform nucleus (transverse section); *C*, inferior temporal (transverse section); *D*, precuneus (sagittal section); *E*, posterior cingulate (sagittal section); and *F*, parietal lobule (sagittal section).

Abnormal Involuntary Movement Scale (AIMS) tests at Tennessee and Mississippi-based nursing homes.

To support his defense, Semrau sought to introduce the expert testimony of Dr. Steven Laken, CEO of one of two corporations that offered fMRI lie detection services at that time (see Box 6.8), and the expert consulted by the plaintiff in the Wilson case. Through his research, Laken concluded that he could apply his fMRI methods to determine deception with 86%–97% accuracy.

According to the court's Admissibility Report and Recommendation, Laken developed 20 neutral questions ("Do you like to swim?" "Are you over age 18?") and 20 control questions ("Do you ever gossip?" "Have you ever done something illegal?") for his test of Semrau. In concert with Semrau's attorney, he also co-developed incident-specific questions ("Did you bill CPT Code 99312 to cheat or defraud Medicare?" "Did you enter into a scheme to defraud the government by billing for AIMS tests conducted by psychiatrists under CPT Code 99301?") related to charges of medication management billing fraud and AIMS billing fraud, respectively.

Laken conducted the two separate scans and concluded that Semrau was "not deceptive" in his answers related to the medication management billing fraud, but that he was "being deceptive" in his responses to the questions related to his AIMS billing practices. Laken indicated that there was only a 6% chance that a positive result on this test was accurate in one

150 Neurocriminology

purporting to tell the truth. This, and the fact that Semrau had reported fatigue following the second test, prompted Laken to administer a modified second test for which the incident-specific questions were shortened to mitigate fatigue.

Based upon the results of the third fMRI, Laken concluded that "Dr. Semrau's brain indicates he is telling the truth in regards to not cheating or defrauding the government." Laken also concluded that, "a finding such as this is 100% accurate in determining truthfulness from a truthful person."

To determine admissibility under Daubert, the court needed to determine if "the methodology, principles, and reasoning are scientifically valid" and "whether the expert's reasoning or methodology can be properly applied to the facts at issue, that is, whether the opinion is relevant."

Based upon a review of Laken's own research, as well as that of others, the court concluded that fMRI for lie detection lacked the known error rates, proven generalizability to real-world scenarios, comparative population norms (Semrau was 63 years of age and most research was conducted on small sample sizes of populations under age 50), and standardized administration protocols necessarily to render it scientifically valid. The court additionally noted that, "fMRI-based lie detection has not yet been accepted by the scientific community."

In addition to its finding under Daubert, the court ruled that the fMRI test in Semrau was inadmissible under Federal Rule of Evidence 403, which requires the court to determine if the unfair prejudice of the evidence substantially outweighs its probative value. The court based its determination, in part, on Laken's responses during cross-examination. Laken conceded that the test results could not indicate whether Semrau responded truthfully to any specific question but only that his collective answers were generally truthful; Semrau could have lied on some of the questions, including some of the incident-specific questions. Laken also acknowledged that the scan results only reflected what Semrau believed at the time of the test, not his mental state was at the time of the actual events.

BOX 6.8 INCORPORATING LIE DETECTION TECHNOLOGY

In 2006, fMRI lie detection was marketed by two companies, No Lie MRI, Inc. and Cephos LLC, for a variety of applications, including employment screenings, drug testing, fraud detection, and even dating. fMRI lie detection was also made available for civil and criminal justice purposes, including a witness's mental state or testimony.

No Lie MRI continues to offer its services for public and private application.

Prevention and Investigation

151

Two years later, in August 2012, a Montgomery County Circuit Court also found fMRI lie detection scans inadmissible. The motion for review was filed by the defense for Gary Smith who was facing a second trial for the murder of fellow former Army Ranger Michael Queen with whom he had served in Afghanistan.

Smith's prior conviction for depraved heart second-degree murder and use of a handgun in the commission of a felony was overturned by a Court of Appeals on the grounds that the prosecution had been allowed to introduce testimony related to Queen's state of mind, while the defense had not been.

Such evidence was relevant, as Smith claimed that Queen's September 2006 death was a suicide.

Smith's account of events was complicated by the multiple versions he provided on the night of the incident. Following his call to 911, police arrived at apartment that he and McQueen shared to find McQueen in the living room, dead from a gunshot wound to the right side of his head. His body was sitting in a chair facing a television, and a bottle of beer, a marijuana bong, and a television remote control were on the floor beside him. No gun was found by McQueen's body or in the apartment.

Police took Smith into custody for questioning. In his initial version of events, he indicated that he and McQueen had gone to several bars before he dropped McQueen at the apartment and drove this mother's house to retrieve some clothes. He stated that when returned, he found McQueen dead in the chair and called 911. He suggested that neither he nor McQueen kept firearms in the apartment.

In the second account, Smith stated that he came home from his mother's house and found McQueen dead with a.38 caliber firearm on the floor next to his hand. He stated that he owned the gun and, realizing that his fingerprints would be on it and the ammunition, panicked. He took the gun to a nearby lake and, after disposing of it, returned to the apartment and dialed 911.

In the third version of events, Smith stated that he took the .38 from his mother's house and placed it on the floor in the living room where McQueen was watching television. He stated that he warned McQueen that the gun was loaded prior to leaving the room to use the bathroom. He stated that he then heard a gunshot, and returned to the living room to find McQueen had shot himself in the head. He stated that it was then that he took the gun and threw it into the lake prior to calling the police.

According to the court's Opinion Order, Smith had undergone both an MRI and an fMRI: "The data for both tests were sent to a company which specializes in analyzing this new technology. The Defendant proposes that, for the first time ever, the results of such an analysis be introduced into a court of law."

Joel Huizenga, CEO of No Lie MRI—the company referenced in the Opinion—submitted supporting documents that stated: "There is always room to do more research in anything, the brain's a complex place. There have been 25 original peer reviewed scientific journal articles, all of them say that the technology works, none of them say that the technology doesn't work…that's 100% agreement."

In contrast, the State presented testimony from professor of psychology and neural science, Dr. Elizabeth Phelps, who stated that only 9% of the 25 papers tendered by the defendant used fMRI for lie detection purposes and that that use of fMRI for lie detection was not yet accepted in their scientific community.

Montgomery County Judge Eric Johnson ruled that "it is clear to the Court that the use of fMRI to detect deception and verify truth in an individual's brain has not achieved general acceptance in the scientific community. Therefore, it does not pass the requisite standard for evidence as delineated in *Frye*…"

In 2012, the U.S. Court of Appeals similarly affirmed a lower court's finding in the case of Semrau. In an interesting footnote, the court added:

> The prospect of introducing fMRI lie detection results into criminal trials is undoubtedly intriguing and, perhaps, a little scary…There may well come a time when the capabilities, reliability, and acceptance of fMRI lie detection—or even a technology not yet envisioned—advances to the point that a trial judge will conclude, as did Dr. Laken in this case: "I would subject myself to this over a jury any day." Though we are not at that point today, we recognize that as science moves forward the balancing of Rule 403 may well lean toward finding that the probative value for some advancing technology is sufficient.

Given the ongoing efforts to translate neuroscience to the criminal justice system since the courts pronouncement in 2012, it is likely that its probative value will be—at the very least—continuously revisited.

Key Terms

Daubert Standard: Standard to determine the admissibility of scientific evidence that has been adopted by many states and the federal courts, and that requires that expert opinion be based upon a scientific technique that meets five elements: whether a method can or has been tested; the known or potential rate of error; whether the methods have been subjected to peer review; whether there are standards controlling the technique's operation; and the general acceptance of the method within the relevant community.

Declarative Memory: The explicit memory for facts and events that can be consciously recalled. Generally, most relevant to testimony.

Diminished Culpability: A legal defense of impaired mental or physical capacity that argues that, although the accused broke the law, they should not be held fully criminally liable for doing so.

Prevention and Investigation

Dopaminergic System: The brain's natural reward circuitry, which has been implicated in drug addiction and abuse.

Downward Departure: A sentence that is less than the recommended statutory minimum.

Frye Standard: Standard to determine the admissibility of scientific evidence that is used in some states and that requires that expert opinion be based upon a scientific technique that is generally accepted as reliable in the relevant scientific community.

Multivoxel Pattern Analysis (MVPA): A statistical analysis that is applied to fMRI data to measure reproducible spatial patterns of activity across experimental conditions.

Neuroplasticity: The brain's ability to adapt, change, and reorganize itself by forming new neural connections throughout life.

rt-fMRI: Real-time fMRI. An imaging methodology that provides subjects with contemporaneous feedback of brain data. Used to support volitional retraining of neural responses.

Use Your Brain

Test Your Knowledge

1. Although the Supreme Court's decision in *Roper v. Simmons* did not specifically reference the neuroscientific research submitted as part of the various amicus briefs, it is credited as a milestone case in considering such evidence. Specifically, the Court concluded that:
   a. Fourteen and fifteen year olds should not be held as criminally responsible as adults and, therefore, should not be eligible for sentences of life without parole.
   b. Juveniles cannot be tried in adult courts given their undeveloped brain functioning.
   c. Sixteen and seventeen year olds cannot be sentenced to capital punishment given the diminished culpability that is associated with their lack of maturity and underdeveloped sense of responsibility.
   d. Sixteen and seventeen year olds should not be held as criminally responsible as adults and, therefore, should not be eligible for sentences of life without parole.
2. The Supreme Court, citing "brain science," determined that youth undergo neurological development that supports rehabilitation and,

154 Neurocriminology

consequently, they cannot be sentenced to life without the possibility of parole, even in murder cases, in which decision:

a. *Stanford v. Kennedy*
b. *Miller v. Alabama*
c. *Graham v. Florida*
d. *Roper v. Simmons*

3. Specialty courts target specific subtypes of criminal behavior and emphasize treatment over retribution in attempts to prevent recidivism. Available research to date on the three major specialty courts—juvenile courts, drug courts, veterans courts—suggest that these courts:

a. Are no more successful than non-specialty courts at preventing recidivism.
b. Are generally more successful than non-specialty courts at preventing recidivism.
c. Are generally less successful than non-specialty courts at preventing recidivism.
d. The research has been inconclusive.

4. To date, studies on witness memory recovery have revealed that subjects can effectively:

a. Feign memories for previously unencountered stimuli
b. Conceal memories for previously encountered stimuli
c. Neither a and b
d. Both a and b

5. To date, the use of fMRI lie detection has received widespread acceptance in the courts during the guilt phase of criminal proceedings.

a. True
b. False

Apply Your Knowledge

1. In recent decades, the Supreme Court has ruled that even when they engage in callous and heinous crimes, youth cannot be held to the same standard of culpability as adults who do likewise. Based upon neuroscientific research reviewed in this and prior chapters, what are some of the arguments that can be made in support of this position? Based upon the neuroscientific research, are there arguments that can be made against this position?

2. Specialty courts target specific criminal subtypes and apply specialized interventions designed to rehabilitate the root cause of the criminal behavior, thereby preventing recidivism. One important aspect of neuroscientific inquiry is identifying the structural and functional

Prevention and Investigation

aspects of the brain that are associated with specific behaviors and, when maladaptive, identifying interventions that might "correct" these behaviors. In what practical ways can neuroscience inform the current specialty courts? In addition to any scientific challenges, are there any socio–cultural–political influences that are relevant when considering the application of neuroscience to any one or all of the specialty courts?

3. Multiple studies have confirmed that trained classifiers can reliably decode multivoxel pattern analysis (MVPA) on fMRI data to determine if cooperative individuals are, in fact, accessing recalled versus novel material. Given the importance and unreliability of eyewitness testimony to many criminal justice proceedings, do you think that the use of this technology should be more commonly adopted? Why or why not?

Answer Key:
1. (c) 2. (b) 3. (b) 4. (d) 5. (b)

Bibliography

Andersen, S. L., & Teicher, M. H. (2009). Desperately driven and no brakes: Developmental stress exposure and subsequent risk for substance abuse. *Neuroscience and Biobehavioral Reviews, 33*, 516–524.

Beauchaine, T., Neuhaus, E., Brenner, S., & Gatzke-Kopp, L. (2008). Ten good reasons to consider biological processes in prevention and intervention research. *Development and Psychopathology, 20*(3), 745–774. doi:10.1017/S0954579408000369.

Bechara, A. (2005). Decision making, impulse control and loss of willpower to resist drugs: A neurocognitive perspective. *Nature Neuroscience, 8*, 1458–1463.

Blair, R. J. (2008). The amygdala and ventromedial prefrontal cortex: Functional contributions and dysfunction in psychopathy. *Philosophical Transactions of the Royal Society of London. Series B, Biological Sciences, 363*, 2557–2565.

Caria, A., Sitaram, R., & Birbaumer, N. (2012). Real-time fMRI: A tool for local brain regulation. *Neuroscientist, 18*, 487–501.

Casey, B., Jones, R. M., & Somerville, L. H. (2011). Braking and accelerating of the adolescent brain. *Journal of Research on Adolescence, 21*, 21–33.

Cohen, A., Bonnie, R., Taylor-Thompson, K., & Casey, B. (2016). When does a juvenile become an adult? Implications for law and policy, *Temple Law Review, 88*, 769.

Coid, J., Yang, M., Roberts, A., Ullrich, S., Moran, P., Bebbington, P., Brugha, T., Jenkins, R., Farrell, M., Lewis, G., & Singleton, N. (2006). Violence and psychiatric morbidity in a national household population: A report from the British Household Survey. *American Journal of Epidemiology, 164*, 1199–1208.

Corrigan, P. W., & Watson, A. C. (2005). Findings from the National Comorbidity Survey on the frequency of violent behavior in individuals with psychiatric disorders. *Psychiatry Research, 136*, 153–162.

deCharms, R. C. (2008). Applications of real-time fMRI. *Nature Reviews Neuroscience, 9*, 720–729.

Diekhof, E. K., Falkai, P., & Gruber, O. (2008). Functional neuroimaging of reward processing and decision-making: A review of aberrant motivational and affective processing in addiction and mood disorders. *Brain Research Reviews, 59*, 164–184.

Ekman, P., O'Sullivan, M., & Friesen, W. V. (1991). Face, voice and body in detecting deceit. *Journal of Nonverbal Behavior, 15*, 125–135.

Elbogen, E. B., Johnson, S. C., Wagner, H. R., Newton, V. M., Timko, C., Vasterling, J. J., & Beckham, J. C. (2012). Protective factors and risk modification of violence in Iraq and Afghanistan War Veterans. *Journal of Clinical Psychiatry, 73*, 767–773. doi:10.4088/JCP.11m07593.

Elbogen, E. B., Johnson, S. C., Wagner, H. R., Sullivan, C., Taft, C. T., & Beckham, J. C. (2014). Violent behaviour and post-traumatic stress disorder in US Iraq and Afghanistan Veterans. *British Journal of Psychiatry, 204*, 368–375. doi:10.1192/bjp.bp.113.134627.

Feldman, D. E. (2009). Synaptic mechanisms for plasticity in neocortex. *Annual Review of Neuroscience, 32*, 33–55.

Fishbein, D., & Tarter, R. (2009). Infusing neuroscience into the study and prevention of drug misuse and co-occurring aggressive behavior. *Substance Use and Misuse, 44*, 1204–1235.

Garrison, K. A., & Potenza, M. N. (2014). Neuroimaging and biomarkers in addiction treatment. *Current Psychiatry Reports, 16*(12), 513. doi:10.1007/s11920-014-0513-5.

Gow, R., Vallee-Tourangeau, F., Crawford, M., Taylor, E., Ghebremeskel, K., Bueno, A., Hibbeln, J., Sumich, A., & Rubia, K. (2013). Omega-3 fatty acids are inversely related to callous and unemotional traits in adolescent boys with attention deficit hyperactivity disorder. *Prostaglandins, Leukotrienes and Essential Fatty Acids, 88*(6), 411–418. doi:10.1016/j.plefa.2013.03.009.

Holst, V., & Pezdek, K. (1992). Scripts for typical crimes and their effects on memory for eyewitness testimony. *Applied Cognitive Psychology, 6*, 573–587.

Hughes, K. C., & Shin, L. M. (2011). Functional neuroimaging studies of post-traumatic stress disorder. *Expert Review of Neurotherapeutics, 11*(2), 275–285. doi:10.1586/ern.10.198.

James, W. (1890, 2017). *The Principles of Psychology, Vols 1 and 2*. San Bernardino, CA: CreateSpace Independent Publishing Platform.

Jordan, K. B., Marmar, C. R., Fairbank, J. A., Schlenger, W. E., Kulka, R. A., Hough, R. L., & Weiss, D. S. (1992). Problems in families of male Vietnam Veterans with posttraumatic stress disorder. *Journal of Consulting and Clinical Psychology, 60*, 916–926. doi:10.1037//0022–006X.60.6.916.

Kassin, S., Ellsworth, P., & Smith, V. (1989). On the "general acceptance" of eyewitness testimony research: A survey of experts. *American Psychologist, 44*, 1089–1098.

Langan, P. A., & Levin, D. J. (2002). *Recidivism of prisoners released in 1994*. Washington, DC: Bureau of Justice Statistics, U.S. Department of Justice.

Laris, M. (2012). Debate on Brain Scans as Lie Detectors Highlighted in Maryland Murder Trial. The Washington Post, Aug. 26.

Prevention and Investigation

MacManus, D., Dean, K., Jones, M., Rona, R. J., Greenberg, N., Hull, L., Fahy, T., Wessely, S., & Fear, N. T. (2013). Violent offending by UK military personnel deployed to Iraq and Afghanistan: a data linkage cohort study. *Lancet, 381*, 907–917. doi:10.1016/S0140–6736(13)60354-2.

McEllistrem, J. E. (2004). Affective and predatory violence: A bimodal classification system of human aggression and violence. *Aggression and Violent Behavior, 10*, 1–30.

Meissner, C., & Brigham, J. (2001). Thirty years of investigating the own-race bias in memory for faces: A meta-analytic review. *Psychology, Public Policy, and Law, 7*:3.

Mitchell, J. (2014). Researchers Find Computer-Based Exercises Significantly Improve the Ability to Pay Attention. Pediatrics, May 14.

Mohamed, F., Faro, S., Gordon, N., Platek, S., Ahmad, H., & Williams, J. M. (2006). Brain mapping of deception and truth telling about an ecologically valid situation: Functional MR imaging and polygraph investigation—initial experience. *Radiology, 238*(2), 679–688.

Perry, J. L., Joseph, J. E., Jiang, Y., Zimmerman, R. S., Kelly, T. H., & Darna, M., et al. (2011). Prefrontal cortex and drug abuse vulnerability: Translation to prevention and treatment interventions. *Brain Research Reviews, 65*, 124–149.

Raine, A., Portnoy, J, Liu, J., Mahoomed, T., & Hibbeln, J. (2015). Reduction in behavior problems with omega-3 supplementation in children aged 8–16 years: a randomized, double-blind, placebo-controlled, stratified, parallel-group trial. *Journal of Child Psychology and Psychiatry, 56*(5), 509. doi:10.1111/jcpp.12314.

Rissman, J., Greely, H. T., & Wagner, A. D. (2010). Detecting individual memories through the neural decoding of memory states and past experience. *Proceedings of the National Academy of Sciences, 107*(21), 9849–9854. doi:10.1073/pnas.1001028107.

Schwartz, I., Steketee, M., & Butts, J. (1991). Business as usual: Juvenile justice during the 1980s. *Notre Dame Journal of Law, Ethics and Public Policy, 5*(2), 377. Retrieved from http://scholarship.law.nd.edu/ndjlepp/vol5/iss2/6.

Semmler, C., Brewer, N., & Wells, G. (2004). Effects of postidentification feedback on eyewitness identification and nonidentification confidence. *Journal of Applied Psychology, 89*, 334–335.

Spohn, C., & Holleran, D. (2002). The effect of imprisonment on recidivism rates of felony offenders: A focus on drug offenders. *Criminology, 40*, 329–357.

Smith v. State, 423 Md. 573 (2011).

Sterzer, P., Stadler, C., Poustka, F., & Kleinschmidt, A. (2007). A structural neural deficit in adolescents with conduct disorder and its association with lack of empathy. *Neuroimage, 37*, 335–342.

Swanson, J. W., Holzer, C. E., Ganju, V. K., & Jono, R. T. (1990). Violence and psychiatric disorder in the community: Evidence from the epidemiologic catchment area surveys. *Hospital and Community Psychiatry, 41*, 761–770.

Uncapher, M. R., TylerBoyd-Meredith, J. T., Chow, T. E., Rissman, J., & Wagner, A. (2015). Goal-directed modulation of neural memory patterns: Implications for fMRI-based memory detection. *The Journal of Neuroscience, 35*(22), 8531–8545.

Wager, T. D., Davidson, M. L., Hughes, B. L., Lindquist, M. A., & Ochsner, K. N. (2008). Prefrontal-subcortical pathways mediating successful emotion regulation. *Neuron, 59*, 1037–1050.

Wilson v. Corestaff Servs L.P. 28 Misc. 3d 425 (2010).

Xue, G., Chen, C., Lu, Z. L., & Dong, Q. (2010). Brain imaging techniques and their applications in decision-making research. *Acta Psychologica Sinica, 42*(1), 120–137.

Neurocriminology in the Criminal Justice System

Prosecution and Sentencing

7

Actus non facit reum nisi mens sit rea
(The act is not cupable unless the mind is guilty)

The contention that an injury can amount to a crime only when inflicted by intention is no provincial or transient notion. It is as universal and persistent in mature systems of law as belief in freedom of the human will and a consequent ability and duty of the normal individual to choose between good and evil.

Supreme Court Justice Robert Jackson, *Morissette v. United States,* **1952**

Learning Objectives

1. Describe the most common ways that neuroscience has been incorporated into the prosecution and sentencing phases of the criminal justice process.
2. Assess the considerations associated with admitting neuroscientific evidence at trial, and the impact of its inclusion on various stages of the process.
3. Analyze the potential, limitations, and appropriateness of neuroscience to informing key considerations in the criminal justice process, such as culpability and risk of future dangerousness.

Introduction

Determining state of mind is critical to rendering justice.

The mind of the accused at time of the offense establishes the moral blameworthiness associated with defining the crime. The mind at time of trial determines whether there is requisite competency to proceed. The mind of the convicted offers insight into future risk of dangerousness, and illuminates additional mitigating or aggravating factors that inform sentencing.

Neuroscience has been used to infer state of mind through the structure and function of the brain, resulting in an exponential increase in its application to

160 Neurocriminology

the prosecution and sentencing phases of the criminal justice process. Various analyses have found its introduction to be most prevalent during the sentencing phase, particularly in capital cases. Neuroscientific evidence has also been increasingly introduced during competency and guilt determinations.

Competency to Stand Trial

In 1960, the Supreme Court rendered a landmark decision-affirming competency as requisite to due process. In *Dusky v. United States* (Box 7.1), the Court concurred with the argument of the solicitor general and determined that, "it is not enough for the district judge to find that 'the defendant [is] oriented to time and place and [has] some recollection of events,' but that the 'test must be whether he has sufficient present ability to consult with his lawyer with a reasonable degree of rational understanding—and whether he has a rational as well as factual understanding of the proceedings against him.'"

The Dusky standard, which the Court subsequently ruled applies to competency to plead guilty and competency to waive the right to counsel, has been generally adopted in most states.

BOX 7.1 THE STANDARDS FOR COMPETENCY

Initial standards for determining competency to stand trial stem from the U.S. Supreme Court decision in *Dusky v. United States.*

The court's ruling involved the case of Milton Richard Dusky, a 33-year-old male who was charged with the 1958 kidnapping of a 15-year-old girl. At Dusky's arraignment, his court-appointed counsel suggested that Dusky might not be competent to stand trial. In response, the court committed Dusky to a forensic medical facility. There, he was evaluated and diagnosed with schizophrenia, which was part of the evidence presented at a subsequent hearing. Despite the diagnosis, Dusky was found to be oriented to time, place, and person—the basic elements of a mental status examination—and deemed competent. He was tried in 1959, pleaded insanity, but found guilty and sentenced to 45 years.

Dusky appealed on several bases, including claims that the trial court erred in finding him competent to stand trial. In its opinion, the Supreme Court ruled that the lower court did not have sufficient information to render a competency decision on Dusky. Specifically, the Court determined that a defendant must be found to have (1) sufficient present ability to consult with his lawyer with a reasonable degree of rational understanding and (2) to have a rational as well as a factual understanding of the proceedings against him.

(Continued)

Prosecution and Sentencing   161

BOX 7.1   THE STANDARDS FOR COMPETENCY (*Continued*)

The Court reversed and remanded.

Subsequently, the Court issued a series of decisions that further refined the standards by which most state and federal jurisdictions determine a defendant's competency.

In its 1975 decision in *Drope v. Missouri*, the Court explored the minimum threshold required for a competency hearing to be warranted. The case involved James Drope, who—along with two others—was arrested in 1969 for the forcible rape of his wife. After severing his case, Drope's attorney requested a continuance so that Drope could undergo the psychiatric evaluation and treatment recommended by a consulting psychiatrist. The motion for continuance was denied. At trial, Drope's wife testified that she had previously not wanted her husband prosecuted but changed her mind after he had "tried to choke me, tried to kill me" prior to the trial. She also testified that her husband had a history of strange behavior, which included rolling himself down stairs when he didn't get his way or was worried about something. On the second day of the trial, Drope was hospitalized after shooting himself in a suicide attempt. The defense's motion for a mistrial due to Drope's absence was denied on the grounds that his absence was voluntary. The jury found Drope guilty and sentenced him to life.

Drope appealed in part on the basis that he was deprived of due process by the failure of the trial court to order a psychiatric examination to determine this competence to stand trial.

The Supreme Court concurred and affirmed that "evidence of a defendant's irrational behavior, his demeanor at trial, and any prior medical opinion on competence to stand trial are all relevant in determining whether further inquiry is required" and that "even one of these factors, standing alone may, in some circumstances, be sufficient" to warrant evaluation. Further, the Court recognized competence as dynamic, concluding that "Even when a defendant is competent at the commencement of his trial, a trial court must always be alert to circumstances suggesting a change that would render the accused unable to meet the standards of competence to stand trial."

In *United States v. Duhon*, the Court provided direction for disposition of cases in which competency will likely not be restored. In July 1997, 20-year-old Keith Joseph Duhon was arrested after picking up photos which depicted his young female cousins in sexually explicit poses. He was charged with sexual exploitation of children and receiving visual depictions of minors engaged in sexually explicit conduct.

(Continued)

BOX 7.1 THE STANDARDS FOR COMPETENCY (*Continued*)

Duhon's attorney filed a motion to suppress statements Duhon made during his arrest, stating that Duhon lacked the mental capacity to waive his Fifth Amendment rights. He also filed an intent to enter an insanity defense.

Several court-appointed mental health professionals concluded that Duhon suffered significantly diminished mental capacity that rendered him incompetent to stand trial. They also determined that he posed a danger to others only if left unsupervised with young children who were also unsupervised. The state moved for Duhon to be placed in a psychiatric mental facility for 4 months in an attempt at competency restoration, despite objections that mental retardation could not be remediated, and concerns that the seizure disorder from which Duhon also suffered could worsen in the confinement condition.

The court ordered a 2-month civil commitment at a Federal Correctional Institution (FCI) at which Duhon participated in a Competency Restoration Group. Thereafter, an FCI psychiatrist and psychologist, who did not conduct psychological evaluations on Duhon, certified him as competent based upon his ability to repeat information he memorized during the competency group. They also reported results of a dangerous assessment screening they conducted, and concluded "there was only a 10% probability that Duhon would commit a sexual offense within the next ten years" and that this small risk could be "significantly reduced with supervision."

Two experts appointed by the Court disagreed with FCI's conclusions regarding Duhon's competency. At a third hearing, Duhon was again found incompetent to stand trial.

In its decision, the Supreme Court provided guidance as to how to manage those not likely to be restored to competency:

> For the foregoing reasons, the undersigned concludes that Duhon is not competent to stand trial and will not attain such capacity in the foreseeable future, with or without hospitalization. The undersigned further concludes that Duhon does not pose a "substantial risk of bodily injury to another person or serious damage to property of another" within the meaning of § 4246. Further hospitalization of the defendant is therefore inappropriate.

The court subsequently expanded the application of the competency standards to other aspects of the criminal justice process—including competency to plead guilty and to waive counsel—in *Godinez v. Moran*.

Prosecution and Sentencing

Two cases related to a single crime were brought before the court in 2008 and 2010 and exemplify the manner in which neuroscientific evidence has contributed to determinations of incompetence as well as competence to stand trial.

During the 2008 competency hearing of Jihad Kasim, a pediatrician accused of insurance fraud, the defense argued that Kasim was incompetent due to a significantly compromised ability to communicate with counsel. To support this assertion, Kasim underwent several neuroimaging scans, including EEGs, an MRI, and SPECT scans. The initial EEG produced abnormal results. A second EEG, as well as the results of an MRI, demonstrated no abnormalities. Kasim was also subject to an SPECT scan, which revealed a reduction in blood flow in the temporal and frontal lobes. Witnesses for the defense testified that these findings signified the presence of frontal lobe dementia, which compromised Kasim's cognitive abilities, including his memory and executive functioning. Despite conflicting opinions regarding the validity of SPECT in the diagnosis of dementia, the court ruled the results admissible, and found that Kasim suffered "an inability to understand the charges against him."

Two years later, the court reached the opposite determination when the same neuroscientific evidence was subject to a different interpretation. Relying upon testimony of state witnesses, the court noted that "...a SPECT scan has been considered an unreliable biological marker of dementia. SPECT measures blood flow in the brain, and that blood flow can change as much as 25% depending on the state of the individual and the resolution of the imaging." The court also noted the significance of Kasim's normal MRI results in contradicting a diagnosis of frontal lobe dementia, a contradiction further substantiated by Kasim's behavior: "A person diagnosed with frontal temporal dementia in 2003, by the end of 2009, would be severely impaired, unable to care for himself, unable to dress himself, incontinent, unable to discern food from other foreign objects, and would exhibit brain changes evident on MRIs and PETs." The court concluded:

> Previously, the expert opinions offered supporting a finding of incompetency both outnumbered and outweighed the single expert offered by the government. However, the closer study of Kasim in the [Board of Prisons (BOP)] medical facility combined with the more accurate medical testing has overcome the prior evidence supporting incompetency. As for the government's contention that Kasim is malingering, the subsequent evidence—most notably the behavior of Kasim when he left the BOP facility and retrieved his test results at the local hospital and when he was recorded during phone conversations—displays a marked difference in Kasim's behavior when he does not think he is being observed, a likely indication of malingering.
>
> The government having met its burden of a preponderance of the evidence with the uncontroverted evidence presented at this hearing, the court finds Kasim competent to stand trial.

164   Neurocriminology

Guilt

Although the lines that define a criminal act are sometimes blurred, it is traditionally and generally held that a crime requires mens rea ("a guilty mind") and actus reus ("a bad act"). An individual may be held civilly liable for bad act without a guilty mind, or feel morally accountable for a guilty mind without a bad act. A finding of criminal responsibility typically requires the intent to do wrong and act of doing so.

In the guilt phase of the criminal process, attempts to introduce neuroscientific evidence are most often made to support a defendant's claim that he or she suffers a brain abnormality that precludes mens rea.

The success of such efforts has been mixed, attributable to both the current state of the science and admissibility standards that require scientific evidence be valid, reliable, and relevant.

BOX 7.2   NEUROIMAGING: PHYSICAL EVIDENCE OR TESTIMONY?

In 1966, the Supreme Court upheld the conviction of a man found guilty of driving under the influence (DUI) of intoxicating liquor. The man was arrested at the hospital where he was receiving treatment for injuries sustained in an automobile accident that occurred during his DUI. At the direction of law enforcement, a blood sample confirming his level of intoxication was taken by medical personnel and introduced as evidence at trial.

The man appealed his conviction on the grounds that the blood draw was taken without his consent, violating his due process rights under the Fourteenth Amendment, his privilege against self-incrimination guaranteed by the Fifth Amendment, his right to counsel under the Sixth Amendment, and his protection against unreasonable search and seizure in violation of the Fourth Amendment.

In *Schmerber v. California*, the Supreme Court upheld the conviction. In its analysis of the claim that a non-consensual blood draw violates the Fifth Amendment privilege against self-incrimination, the Court found that:

> the privilege protects an accused only from being compelled to testify against himself, or otherwise provide the State with evidence of a testimonial or communicative nature, and that the withdrawal of blood and use of the analysis in question in this case did not involve compulsion to these ends.

Citing its 1910 finding in *Holt v. United States*, the Court reiterated that a "[T]he prohibition of compelling a man in a criminal court to be

(Continued)

Prosecution and Sentencing

BOX 7.2   NEUROIMAGING: PHYSICAL EVIDENCE OR TESTIMONY? (*Continued*)

witness against himself is a prohibition of the use of physical or moral compulsion to extort communications from him, not an exclusion of his body as evidence when it may be material."

The testimony versus physical evidence dichotomy adopted by the courts raises potentially interesting questions for the future of neuroscientific evidence.

Under current standards, it would be rational for courts to accept, for example, the introduction of structural MRI data that demonstrates brain damage consistent with injuries that an accused sustained during the commission of a crime—akin to the admission of evidence related to bite marks or scratches. Conversely (and provided such evidence could meet reliability and validity standards), neuroscientific evidence that confirms memory of criminal engagement is more equivocal. Some have suggested that such evidence would and should be treated like other physical evidence particular to the individual, like DNA or fingerprints, and therefore be admissible if lawfully obtained. Others see a distinction between cognition and other forms of evidence that might be "extracted" from the body; in addition to suggesting the possible implication of the Sixth Amendment protection against unreasonable search and seizure, being compelled to share the silent utterances of the mind could be argued as violating the Fifth Amendment privilege against self-incrimination.

Given its potentially broad applications both within the criminal justice system, and in relation to national and private security, there has been a significant investment in the science of "reading minds." Presuming this investment yields valid, reliable, and relevant results that are admissible in criminal proceedings, the courts of the future may be forced to a more nuanced interpretation of the distinction between testimony and physical evidence.

Challenges to validity and reliability are illustrated in the court's decision to deny the introduction of PET scans in the highly publicized case of *People v. Goldstein*. In January 1999, 32-year-old Andrew Goldstein pushed 32-year-old Kendra Webdale from a New York City subway platform into an oncoming train, killing her. Goldstein, who was diagnosed as schizophrenic, sought to introduce a PET scan that showed his brain had a "massive reduction in metabolism in the frontal lobe and the basal ganglia," which the defense argued offered proof of his schizophrenia.

166   Neurocriminology

The diagnosis of schizophrenia was relevant to the defense's case that Goldstein was not responsible for his act due to his mental illness, for which he had stopped taking medication, rendering him incapable of understanding his actions and whether they were right or wrong.

The PET scan evidence was precluded.

Goldstein's first trial ended in a hung jury. During his second, the jury rejected his insanity defense and found him guilty of second-degree murder.

The failure of the court to introduce the PET scan evidence was one of the bases for his appeal. Although the court granted a new trial, it rejected the argument that the PET evidence should have been included. The court concluded that evidence lacked both validity and relevance:

> The record establishes that defendant's PET scan evidence was properly precluded. The prosecution had conceded defendant's mental illness, and their expert witnesses had testified that defendant was schizophrenic. Moreover, the special master appointed by the court stated that the test results could not conclusively prove schizophrenia, but could only show an abnormality in the brain, which in either case would not be probative of the key issue of the insanity defense, i.e., whether defendant comprehended either the nature and consequences of his actions or that his actions were wrong, since a diagnosis of schizophrenia does not preclude per se that defendant is capable of such comprehension.

In October 2006, Goldstein pleaded guilty to manslaughter, and was sentenced to 23 years in prison with 5 years of post-release supervision, including psychiatric oversight. As part of his plea agreement, Goldstein admitted to intentionally pushing Kendra from the subway platform:

She was leaning against a pole with her back to me near the edge of the platform by the tracks. I looked to see if the train was coming down the tracks. I saw that the subway train was coming into the station. When the train was almost in front of us, I placed my hands on the back of her shoulders and pushed her. My actions caused her to fall onto the tracks.

BOX 7.3   KENDRA'S LAW

Following the murder of Kendra Webdale, it was determined that murderer Andrew Goldstein had a history of short-term psychiatric treatment. He also had a history of violence that included shoving and hitting people.

Absent incarceration or placement in a long-term psychiatric facility, there was no mechanism or means to ensure that Goldstein obtained the treatment needed to support his stability.

In 1999, New York passed Kendra's Law. Under Kendra's Law, courts are permitted to mandate that certain individuals with serious mental

(Continued)

Prosecution and Sentencing

BOX 7.3 KENDRA'S LAW (*Continued*)

illness undergo publically-subsidized, community-based assisted outpatient treatment (AOT) for up to 1 year. Individuals subject to Kendra's Law must have a history of multiple incidents of homelessness, arrests, incarceration, and/or hospitalization.

In addition to a finding of severe mental illness, the court must determine that the individual meets the following criteria:

1. Is unlikely to survive safely in the community without supervision
2. Has a history of non-compliance with treatment that has:
   a. been a significant factor in his or her being in a hospital, prison or jail at least twice within the last 36 months or
   b. resulted in one or more acts, attempts or threats of serious violent behavior toward self or others within the last 48 months
3. Is unlikely to voluntarily participate in treatment
4. Is, in view of his or her treatment history and current behavior, in need of AOT in order to prevent a relapse or deterioration which would be likely to result in:
   a. a substantial risk of physical harm to the consumer as manifested by threats of or attempts at suicide or serious bodily harm or conduct demonstrating that the consumer is dangerous to himself or herself, or
   b. a substantial risk of physical harm to other persons as manifested by homicidal or other violent behavior by which others are placed in reasonable fear of serious physical harm.

To date, two studies have been conducted to evaluate the efficacy of Kendra's Law on several key outcome measures: 74% fewer experienced homelessness, 77% fewer experienced psychiatric hospitalization; 83% fewer experienced arrest and 87% fewer experienced incarceration than in the 3 years prior to participation. Additionally arrest rates decreased from 3.7% to 1.9% in the month following participation, and hospitalization rates declined from 74% to 36% in the 3-month follow-up period.

The overall decrease in harmful human behavior for those who participated was also significant at 44%.

Of particular interest: The mandated aspect of treatment was associated with better outcomes. When compared to those receiving

(Continued)

168   Neurocriminology

> ### BOX 7.3   KENDRA'S LAW (*Continued*)
>
> identical levels of service on a voluntary basis, those receiving service under court order were significantly less likely to be hospitalized (36% versus 58%), and less likely to be arrested (1.9% arrested per month versus 2.8%).
>
> At least in relation to the targeted population, the outcomes suggest the combined influence of the criminal justice system and science is more impactful than either alone.

Reliability has also played a factor in the court's decision-making related to the introduction of neuroscientific evidence, as illustrated in the case of *U.S. v. Montgomery*. In *Montgomery*, the court considered an appeal that stemmed from the case of Lisa Montgomery who, in 2007 was convicted of kidnapping resulting in death, for which she received a death penalty verdict.

In December 2004, Montgomery assumed an alias and expressed an interest in purchasing a puppy from a woman with whom she had become acquainted at a dog show the preceding April. The two kept in touch through online message boards regarding dog breeding. The woman announced her pregnancy online. Shortly thereafter, Montgomery likewise told her husband, family, and the online community that she was pregnant—despite the fact that she had been sterilized years earlier.

The woman was 8 months pregnant when Montgomery used the ruse of purchasing the puppy to meet her. Instead, she attacked the woman with a kitchen knife and strangled her with a cord, each of which she had brought with her. She then used the knife to cut into the woman's abdomen, causing her to regain consciousness. Montgomery strangled her a second time, killing her. She extracted the fetus from the woman's body, cut the umbilical cord, and left with the physically unharmed baby.

Montgomery called her husband from the road and informed him that she had given birth. The two returned to their home and announced the birth to friends and family.

The following day, law enforcement questioned Montgomery and, after initially lying about the circumstances of the child's birth, she confessed to the murder, and to kidnapping the child.

In her appeal, Montgomery claimed that she could not be convicted of kidnapping because Stinnett died before the baby attained legal personhood by being born. In addition, Montgomery asserted that the trial court was in error in denying the admissibility of a PET scan and MRI that demonstrated

Prosecution and Sentencing

that she had structural and functional brain abnormalities consistent with psuedocyesis, or false pregnancy. Psuedocyesis is rare and most often diagnosed in women in developing countries, particularly when there is strong socio-cultural pressure to be fertile. Clinically, pseudocyesis is differentiated from delusions of pregnancy that are often a sign of psychosis; pseudocyesis is considered a somatoform disorder and is frequently accompanied by physical signs of pregnancy including abdominal swelling, menstrual disturbance, breast tenderness, and weight gain.

Montgomery's PET scan revealed elevated activity throughout the limbic system, and in the somatomotor region, including the hypothalamus. During the in camera—or private, in chamber—Dalbert hearing, the defense witness testified that "heightened activity in the hypothalamus has been shown to produce pseudopregnancy in rats" and was consistent with the diagnosis of pseudocyesis: "the brain she has may explain at least part of what happened."

Two prosecution witnesses countered that the PET scan results could not be used to identify or diagnosis specific conditions, including pseudocyesis. One expert asserted that the scan "is not particularly relevant or important to say it's consistent with pseudocyesis. It's consistent with many, many, many things."

Both prosecution witnesses also questioned the defense expert's methodology, claiming that he calculated the data from the control group differently than he did for the data obtained from Montgomery's scan. Consequently, the results were unreliable; the experts were only able to reproduce the defense witness's calculations if they likewise used the two different calculation methods. The witnesses for the prosecution also expressed concerns over testing conditions and scanner resolution.

Following the hearing, the court ruled that it would admit the PET evidence on a limited basis: The defense could testify regarding the PET findings of abnormalities in the somatomotor region of Montgomery's brain, and the association between these abnormalities and pseudocyesis. The defense could not testify, however, that these abnormalities had a causal effect on Montgomery's behavior, nor that the scan—which was taken 3 years after the index offense—was consistent with Montgomery's brain functioning at the time of the event.

Prior to opening statements, the court decided to preclude even the limited admissibility of the PET scan results when the defense expert failed to produce the original data obtained from the control group, thereby preventing the prosecution from duplicating the results without applying an invalid methodology. The resulting lack of reliability, which would inevitably lead to courtroom disputes over calculations, was deemed to be likely more confusing than probative.

170   Neurocriminology

Not all courts have rejected neuroscientific evidence during the guilt phase.

In 2006, the U.S. Court of Appeals in *U.S. v. Sandoval-Mendoza*, for example, reversed a felony conviction on the basis of a lower court's exclusion of neuroscientific evidence. Eduardo Sandoval-Mendoza and his twin brother were convicted of conspiring to sell methamphetamine. As part of his appeal, Eduardo argued that the court erred in excluding evidence of a brain tumor that he argued made him particularly vulnerable to entrapment by law enforcement.

Eduardo was introduced by a family friend to another individual to whom he sold approximately 12 pounds of methamphetamine in three separate drug deals. Both the friend and the drug buyer were government informants.

Sandoval-Mendoza admitted to selling the drugs. However, he claimed that he was entrapped as the government informants knew that he suffered from a large brain tumor, which "rendered him especially susceptible to suggestion." Specifically, Sandoval-Mendoza testified that he was depressed, and worried about dying and providing for his wife and five children. He spoke of his concerns with the government informant who, he testified, told him that the could earn $5,000–$10,000 a week selling drugs, which he could save to support his family after he died. After refusing to sell drugs for several months, he reported that he eventually capitulated and made the three drug sales for which he was arrested.

To support his defense, Sandoval-Mendoza sought to introduce the testimony of two expert witnesses, a psychologist and a neurologist. The former would testify that the damage caused by the tumor to Sandoval-Mendoza's frontal lobe, temporal lobe, and thalamus "may affect memory, decision-making, judgment, mental flexibility, and overall intellectual capacity" and "in particular, damage to the frontal lobe often affects concentration, focus, learning, memory, decision-making, reasoning, judgment, and problem-solving."

The neurologist who conducted an MRI would testify that Sandoval-Mendoza suffered from an unusually large pituitary tumor that, when shrank by medication, caused the frontal lobe to herniate into the empty space, the inside of the left temporal lobe to atrophy, and penetrated a bone separating the pituitary gland from the brain stem. The resulting brain damage would affect intellectual functioning, including judgment, memory, and emotions connected to memory.

During the in camera Daubert hearing, both experts testified that they were unaware of research specifically linking the type of brain damage from which Sandoval-Mendoza suffered with susceptibility to commit crimes. Additionally, a neurologist retained as expert witness for the prosecution testified that there was insufficient research to conclude that the tumor and

Prosecution and Sentencing   171

brain damage impacted Sandoval-Mendoza's behavior and cognition. Citing two of the generally recognized limitations of neuroscientific evidence, the prosecution witness argued that research which served as the bases for inferring a link between Sandoval-Mendoza's brain tumor and his behavior was invalid as subjects were studied retrospectively—the brains of research subjects were not observed at the time of the act—and the sample sizes involved were small. To further accentuate the distinction between correlation and causation, the prosecution witness referenced a patient of his who had a tumor like Sandoval-Mendoza's, and who did not commit crimes.

The court concluded that the defense's expert testimony was "not relevant to the entrapment defense" because it "does not tend to show either inducement or a lack of predisposition attributable to the tumor." In addition to concluding that the testimony lacked scientific validity and failed to make a "causal connection" between the tumor and the criminal behavior, the court determined that any probative value would be "outweighed by the dangers of confusing the issues, misleading the jury, and creating undue delay."

Sandoval-Mendoza was tried, convicted, and sentenced to 235 months.

In reversing his conviction, the Court of Appeals noted the following:

> The district court concluded that the proposed medical expert opinion testimony was unreliable because it did not conclusively prove Sandoval-Mendoza's brain tumor caused susceptibility to inducement or a lack of predisposition. But medical knowledge is often uncertain. The human body is complex, etiology is often uncertain, and ethical concerns often prevent double-blind studies calculated to establish statistical proof. This does not preclude the introduction of medical expert opinion testimony when medical knowledge "permits the assertion of a reasonable opinion"... And there was no risk of "misleading the jury." The experts agreed Sandoval-Mendoza has an unusually large brain tumor. Their only disagreement was whether it caused susceptibility to inducement. The jury was capable of weighing the conflicting medical expert opinion testimony against the rest of the evidence presented and determining whether or not predisposition existed. As for "undue delay," testimony would likely consume no more time than the Daubert hearing, and probably much less.
>
> Without the medical expert opinion testimony, the real issue in dispute was hidden from the jury. It could not determine whether the government's informants induced a vulnerable and suggestible man to break the law. The informants did not testify, so the jury could not evaluate the pressure they put on Sandoval-Mendoza. It could not evaluate the merits of Sandoval-Mendoza's suggestibility, because the medical expert opinion testimony concerning the possibility his tumor or limited mental capacity made him susceptible to inducement was excluded. All the jury had was proof that Sandoval-Mendoza sold drugs, wiretap recordings in which he sounded like an experienced drug dealer, and a couple of lay witnesses testifying that he was addled by a brain tumor. Sandoval-Mendoza is entitled to present his case to the jury. For that, he deserves a new trial.

172   Neurocriminology

Sentence Mitigation

Once a judge or jury finds a defendant guilty of a crime, or the defendant pleads guilty, the penalty phase begins. The rules of evidence admissibility in effect during the guilt phase do not apply during the penalty phase. Instead, state and federal criminal statutes often set maximum penalties based on offense type. Within this context, there is discretion as to length of sentence, and prosecutors and defendants have the opportunity to present evidence related to aggravating and mitigating factors.

Generally, aggravating factors include criminal history, victim type, the heinousness of the crime, and risk of future dangerousness. For capital cases, the Supreme Court has issued a series of decisions that explicate the constitutional rights of defendants to present mitigating evidence, which have also been adopted by many state and federal courts in deciding sentencing for both capital and non-capital cases (see Box 7.4). Factors considered mitigating include a lack of a prior criminal record, a minor role in the offense, the culpability of the victim, a history of abuse or neglect, circumstances at the time of offense, remorse, physical illness, or mental illness and impairment. For the last, the Supreme Court has explicitly ruled that cognitive or neuropsychological impairment may be considered a mitigating factor even if not directly relevant to the crime.

BOX 7.4   THE CONSTITUTIONAL RIGHT TO MITIGATE PENALTIES FOR CRIMINAL BEHAVIOR

Neuroscientific evidence is most often introduced during the penalty phase of criminal trials. This results, in large part, from three 1970s rulings related to a defendant's right to enter mitigating evidence during this stage of his or her defense.

FURMAN V. GEORGIA

Bifurcated guilt-penalty determinations can be traced to the Supreme Court's 1972 decision in *Furman v. Georgia*. After joining three death penalty appeals—one in which defendant Furman was sentenced to death for murder, and two in which defendants Jackson and Branch were independently sentenced to death for rape convictions—the Court determined that capital penalties as then operationalized violated Eighth Amendment protections against cruel and unusual punishment. Although not unanimous, part of the Justices reasoning
(Continued)

Prosecution and Sentencing

BOX 7.4   THE CONSTITUTIONAL RIGHT TO MITIGATE PENALTIES FOR CRIMINAL BEHAVIOR (*Continued*)

resulted from an analysis that racial bias played a role in death penalty convictions:

> The high service rendered by the "cruel and unusual" punishment clause of the Eighth Amendment is to require legislatures to write penal laws that are evenhanded, nonselective, and nonarbitrary, and to require judges to see to it that general laws are not applied sparsely, selectively, and spottily to unpopular groups.

GREGG V. GEORGIA

To comply with the Court ruling, several states enacted legislation that supported the two-phase trial, offered guidelines for imposing sentencing. Georgia was one such state.

In 1976 in *Gregg v. Georgia*, the Court indicated that this approach could satisfy prior Constitutional concerns.

Troy Gregg was convicted of committing armed robbery and murder. At the penalty stage, the judge instructed the jury that it could recommend either a death sentence or a life prison sentence on each count. The judge also instructed the jury that it could consider mitigating or aggravating circumstances and that it would not be authorized to consider imposing the death sentence unless it first found beyond a reasonable doubt (1) that the murder was committed while the offender was engaged in the commission of other capital felonies, viz., the armed robberies of the victims; (2) that he committed the murder for the purpose of receiving the victims' money and automobile; or (3) that the murder was "outrageously and wantonly vile, horrible and inhuman' in that it 'involved the depravity of [the] mind of the defendant."

The jury found in favor of conditions 1 and 2 and sentenced Gregg to death.

In their opinion upholding the death penalty sentence, the Supreme Court including the following:

> The new procedures on their face satisfy the concerns of Furman, since before the death penalty can be imposed there must be specific jury findings as to the circumstances of the crime or the character of the defendant, and the State Supreme Court thereafter reviews the comparability of each death sentence with the sentences imposed on similarly situated defendants to ensure that the sentence of death in a particular case is not disproportionate.

(Continued)

BOX 7.4 THE CONSTITUTIONAL RIGHT TO MITIGATE PENALTIES FOR CRIMINAL BEHAVIOR (*Continued*)

LOCKETT V. OHIO

In the 1978 case of *Lockett v. Ohio*, the Supreme Court more directly articulated the requirement that mitigating evidence be introduced in capital cases:

> The Eighth and Fourteenth Amendments require that the sentencer, in all but the rarest kind of capital case, not be precluded from considering as a mitigating factor, any aspect of a defendant's character or record and any of the circumstances of the offense that the defendant proffers as a basis for a sentence less than death.

On this basis, the Court remanded the decision of a lower court, which had sentenced Sandra Lockett to death after she was found guilty of aggravated murder for her role in planning and driving the car from the scene of a pawn shop robbery during which the shop owner was killed. Under Ohio law at the time, the judge was limited to considering three mitigating factors: (1) whether the victim induced or facilitated the offense; (2) whether it was unlikely that the offense would have been committed but for the fact that the offender was under duress, coercion, or strong provocation; or (3) whether the offense was primarily the product of the offender's psychosis or mental deficiency.

The Supreme Court noted that "The Ohio death penalty statute does not permit the type of individualized consideration of mitigating factors required by the Eighth and Fourteenth Amendments" finding:

> A statute that prevents the sentencer in capital cases from giving independent mitigating weight to aspects of the defendant's character and record and to the circumstances of the offense proffered in mitigation creates the risk that the death penalty will be imposed in spite of factors that may call for a less severe penalty, and, when the choice is between life and death, such risk is unacceptable and incompatible with the commands of the Eighth and Fourteenth Amendments.

Consequently, neuroscientific evidence has increasingly been introduced during the penalty phase of the criminal justice process as a means to prove the presence of a potentially mitigating brain abnormality or dysfunction. The introduction of such evidence is reinforced by the Supreme Court's 1984 decision in *Strickland v. Washington*, through which the court formalized the standards by which a defendant can claim a violation of the Sixth Amendment right to effective assistance of counsel; failure of a defense attorney to investigate, acquire, or produce neuroscientific evidence is increasingly

Prosecution and Sentencing

cited in cases in which Strickland claims are proffered. The introduction of this evidence has prevailed despite concerns that the same evidence introduced to mitigate a defendant's culpability could be usurped by the prosecution to argue intractable dangerousness—what has been referred to as the "double-edged sword" problem (see Box 7.5).

BOX 7.5 THE DOUBLE-EDGED SWORD

On the morning of October 25, 1979, Johnny Penry brutally raped, beat, and stabbed a woman in her Texas home. The victim was able to give a description of Penry prior to dying during the course of emergency treatment. Based upon her description, local sheriff deputies apprehended Penry, who had recently been released on parole after conviction for another rape. Penry confessed to the crime and was subsequently convicted of capital murder.

During the competency hearing and guilt–innocence phase of his trial, a psychologist testified that Penry was mildly to moderately retarded and had the mental age of a 6.5 year old. Evidence also indicated that he had been abused as a child.

At the penalty phase of the trial, Penry offered mitigating evidence of his mental retardation and abused childhood as the basis for a sentence of life imprisonment rather than death. However, per instructions by the court, the jury was only permitted to make its sentencing determination based upon answers to three questions: Did Penry act deliberately when he murdered his victim? Was there a probability that he would be dangerous in the future? Did he act unreasonably in response to provocation?

The jury was never instructed that it could consider Penry's mental retardation or history of abuse as mitigating factors that could inform their sentencing decision, and potentially support a determination that Penry should receive a sentence of life imprisonment.

Instead, the jury answered yes to each question posed and, as required by Texas law, the court sentenced Penry to death.

In its opinion to remand the case for re-sentencing due to a failure of the court to allow mitigating factors to be presented for jury consideration, the Supreme Court noted the following:

> The mitigating evidence concerning Penry's mental retardation indicated that one effect of his retardation is his inability to learn from his mistakes. Although this evidence is relevant to the second issue, it is relevant only as an aggravating factor because it suggests a "yes" answer to the question of future dangerousness. The prosecutor argued at the penalty hearing that

(Continued)

176   Neurocriminology

> **BOX 7.5   THE DOUBLE-EDGED SWORD (*Continued*)**
>
> there was "a very strong probability, based on the history of this defen-
> dant, his previous criminal record, and the psychiatric testimony that
> we've had in this case, that the defendant will continue to commit acts of
> this nature." Even in a prison setting, the prosecutor argued, Penry could
> hurt doctors, nurses, librarians, or teachers who worked in the prison.
>
> Penry's mental retardation and history of abuse is thus a two-edged
> sword: it may diminish his blameworthiness for his crime even as it indi-
> cates that there is a probability that he will be dangerous in the future.
>
> Despite the prosecution's argument, the Court—citing *California v.
> Brown*—ruled that to make an assessment as to whether death repre-
> sented an appropriate sentence, evidence about the defendant's back-
> ground and character is relevant because of the belief, long held by this
> society, that defendants who commit criminal acts that are attributable
> to a disadvantaged background, or to emotional and mental problems,
> may be less culpable than defendants who have no such excuse.
>
> Research has suggested that concerns for the two-edge sword (or
> double-edged sword, as it is sometimes referred) have proven unwar-
> ranted. When introduced at trial during the sentencing phase, neu-
> roscientific evidence is predominantly used to evaluate a defendant's
> moral culpability, rather than used against him or her as evidence of
> a dysfunction that increases the probability of future dangerousness.

The increasing expectation that neuroscientific evidence should be introduced during the penalty phase has not been met by its proportional success in reducing defendant sentences. As in other stages of the process, the
translation of neuroscience to sentencing has had mixed results, illustrated
by several early cases.

The Sentencing of Brian Dugan

In November 2009, an Illinois court held a Frye hearing to determine the
admissibility of fMRI evidence in the penalty phase of the trial of Brian Dugan,
who pleaded guilty to the murder of a 10-year-old girl. Dugan had been serving two life sentences for the 1984 rape and murder of a 27-year-old nurse, and
the 1985 rape and murder of a 7 year old. When apprehended for those crimes,
Dugan had also offered to plead guilty to the 1983 rape and murder of the 10
year old in exchange for the prosecution's agreement not to pursue the death
penalty. The prosecution declined, in part because two others had been convicted and were on death row for the murder of the 10 year old. When the two
were later exonerated, the state used DNA evidence to tie Dugan to the murder.

Prosecution and Sentencing

The defense retained neuroscientist Kent Kiehl to conduct an examination of Dugan. Kiehl is a leading researcher in the study of brain correlates to psychopathy and the use of neuroscience in the prediction of future dangerousness (see Box 7.6). The results of Dugan's fMRI revealed brain abnormalities consistent with psychopathy. Given the impulsivity, lack of remorse, and antisocial behavior that are characteristic of the illness, psychopathy could be considered a mitigating factor in Dugan's defense.

BOX 7.6   THE NEUROPREDICTION OF DANGEROUSNESS

Kent Kiehl, Professor of Psychology, Neuroscience and Law at the University of New Mexico, is a leading neuroscientific researcher in the area of psychopathy. In several studies, Kiehl has found associations between limbic and paralimbic brain region abnormalities and psychopathy.

In a large-scale study published in 2018, he and researchers Flor Espinoza, Victor Vergara, Daisy Reyes, Nathaniel Anderson, Carla Harenski, Jean Decety, Srinivas Rachakonda, Eswar Damaraju, Barnaly Rashid, Robyn Miller, Michael Koenigs, David Kosson, Keith Harenski, and Vince Calhoun conducted an analysis of the resting state fMRI scans from 985 incarcerated adult males aged 18–63. The subjects were assessed for psychopathy using the Psychopathy Checklist–Revised (PCL-R), a standardized psychological assessment tool developed by Dr. Robert Hare to evaluate the presence of psychopathy in forensic populations. The assessment is based upon two correlated factors. Factor 1 represents affective and interpersonal traits, such as callousness and lack of remorse. Factor 2 represents lifestyle and antisocial traits, including impulsivity and early behavior problems. A score of 30 or above is typically associated with the diagnosis of pyschopathy. Of the 985 inmates studied by the researchers, 173 met this threshold. The study revealed functional reduction in connectivity between the amygdala and the cuneus, which is located in the occipital lobe and is associated with self-referential processing, and those who scored high on Psychopathy Factor 1.

Psychopathy is one of the factors incorporated in standardized violence risk assessments. An additional study by Kiehl and fellow researchers Eyal Aharoni, Gina Vincent, Carla Harenski, Vince Calhoun, Walter Sinnott-Armstrong, and Michael Gazzaniga demonstrated a potential neurocorrelate of psychopathy that might aid in the prediction of recidivism. The researchers focused on the anterior cingulate cortex (ACC) region of the limbic area, which is associated

(Continued)

> ## BOX 7.6 THE NEUROPREDICTION OF DANGEROUSNESS (*Continued*)
>
> with emotional regulation, decision-making, conflict negotiation, and avoidance learning. Reduced ACC activity has been correlated with aggression, dysregulation, and apathy.
>
> Kiehl has also found lower ACC activity to be associated with significantly higher errors of commission—falsely responding to stimuli during the "no-go" trials of go-no-go tests which signals a lack of inhibitory control—while undergoing fMRI scans.
>
> Applying this finding, he and his fellow researchers hypothesized that reduced ACC activity would be associated with higher re-arrest rates in a study sample of 96 adult offenders. In reviewing re-arrest records for up to 4 years following incarceration release, the researchers found a significant correlation between lower ACC levels and re-arrest rates when all types of offenses were considered.
>
> Re-arrest rates were too small assess for a specific correlation between low ACC activity and violence. However, the research findings suggest an area for additional study that could supplement current assessments of violence risk and—if able to be replicated—could meet the reliability standards necessary for admissibility in court, helping to inform decision-making related to future dangerousness. Perhaps more importantly, such research might support the development targeted treatment interventions that could prevent violence recidivism.

Kiehl did not claim that brain abnormality caused Dugan to commit his crimes. Experts for the prosecution also stressed the inability to make a causal link between Dugan's fMRI results and his criminal behavior. Prosecution witnesses also protested the lack of relevance of the fMRI, akin to claims frequently cited in relation to neuroscientific evidence during the guilt phase of trials: The presence of a brain abnormality so long after the commission of a crime—in the case of Dugan, 26 years following his commission of the rape and murder—was irrelevant to Dugan's state of mind at the time of the crime's commission.

In what was widely reported to be the first time, a judge allowed testimony related to fMRI evidence to be admitted during the penalty phase of a trial. The judge ruled, however, that Kiehl was not permitted to show the actual brain scans. Instead, he testified using cartoon-like depictions to explain his findings.

On November 10, 2009, the jury deliberated for 5 hours before reporting that, although they had reached a verdict, they wanted additional time. The judge granted them an additional day. On November 11, they sentenced Brian Dugan to death.

Prosecution and Sentencing   179

Media reports following the conclusion of the deliberations suggested that the jury was initially going to recommend a sentence of life in prison, as one juror would have voted against the death penalty—a sentence that requires a unanimous vote. Prior to reading the sentence publically, however, the juror changed votes.

Although it did not ultimately affect the outcome, the fMRI evidence was widely credited with the duration and vacillation that accompanied the jury's deliberation.

Dugan's sentence was ultimately commuted to life without parole in 2011, when Illinois abolished the death penalty.

A year after the jury in the Dugan case rejected neuroscientific evidence as proof of a mitigating element, another jury would cite it as the specific rationale for reaching the opposite decision.

State of Florida v. Grady L. Nelson

In January 2005, following release from a Florida prison for the rape of his stepdaughter, 53-year-old Grady Nelson stabbed his wife and two stepchildren. His wife died as a result of the particularly vicious attack during which Nelson inflicted 60 stab wounds. The children survived the attack. Nelson confessed to the murders, a confession that his attorneys would later argue was coerced.

In 2010, a jury convicted Nelson of first-degree murder for the killing of his wife, a crime for which he faced the death penalty. He was also convicted of sexual battery of his stepdaughter and two counts of attempted first-degree murder.

As a mitigating factor, the defense sought admission of quantitative EEG (QEEG) data, which revealed that Nelson suffered frontal lobe damage consistent with a history of traumatic brain injury, early childhood abuse, and prenatal alcohol exposure. The damage, the defense would assert, demonstrated that Nelson suffered a brain-based predisposition to impulsivity, dysregulation, and violence, which it argued was a mitigating factor.

After conducting a Frye hearing, the judge concluded that "(E)verything I have heard, the methodologies are sound, the techniques are sound, the science is sound."

After only an hour's deliberation, the jury decided to sentence Nelson to life. In media interviews following the deliberation, two jurors cited the neuroscientific evidence as influential in their decision-making: "...the facts about the QEEG, some of us changed our mind" and "It turned my decision all the way around. The technology really swayed me... After seeing the brain scans, I was convinced this guy had some sort of brain problem."

A third juror reported that he also voted for life in prison, rather than the death penalty. His reasoning was unrelated to the QEEG data. Rather,

180 Neurocriminology

news sources stated that, "he wanted Nelson to rot in prison with the stigma of being a child rapist."

Ineffective Use of Counsel

In the span of a few decades, the introduction of neuroscientific evidence at the sentencing phase of the criminal justice process has gone from controversial to commonplace to, increasingly, required. Although the vast number of Strickland claims fail, success is bolstered when counsel cannot articulate a compelling strategy for failure to introduce mitigating evidence, and that mitigating evidence is neuroscientific. In her comprehensive Neuroscience Study, Dr. Deborah Denno, Founding Director of the Neuroscience and Law Center at Fordham University School of Law, found that when Strickland claims were raised in relation to the more than 500 1992–2012 cases in which neuroscientific evidence was introduced, nearly all successful claims resulted from the defense attorney's failure to obtain, investigate, or understand neuroscientific evidence.

Anderson v. Sirmons exemplifies the court's decision-making in such cases.

Glenn Douglas Anderson v. Marty Sirmons

In 1999, a jury convicted Glenn Douglas Anderson of three counts of first-degree murder stemming from the deaths of three acquaintances who died as a result of gunshot wounds or fire-related injuries following an altercation with Anderson and two associates. Anderson was subsequently sentenced to death for each of the three murder convictions. After a series of denied appeals, Anderson challenged the validity of his death sentence on grounds of ineffective assistance of counsel.

In consideration of the appeal, the court summarized the aggravating elements put forth by the prosecution: The "callous and brutal" nature of the crimes. Anderson's post-incarceration behavior, which included drug and knife possession. The likelihood that Anderson would commit future acts of violence. The fact that two of the murders were committed in an attempt to avoid arrest or prosecution.

In contrast, the defense's mitigating evidence consisted of testimony of Anderson's family and coworkers, describing him as "a kind, hard-working, normal man who could be of some help to his daughter if his life were spared."

The court found that the mitigating evidence "played into the prosecution's theory that the only explanation for the murders was that Anderson was simply an 'evil' man."

The court went on to explicitly detail the mitigating evidence it expected of defense counsel in presenting its case on Anderson's behalf:

Prosecution and Sentencing

... the prosecution was able to argue convincingly to the jury that there was nothing in the case to diminish Anderson's moral culpability for the murders. The evidence developed by habeas counsel demonstrates Anderson suffers from brain damage; is "borderline mentally defective"; and functions below the bottom two percent of the general population. Anderson was only able to complete the eighth grade of school. The most significant damage to Anderson's brain is in the area of the frontal lobe, the area of the brain that affects reasoning, problem solving, and judgment. Anderson has suffered chronic drug addiction, which addiction began at the age of nine with the use of alcohol, marijuana, and inhalants and ultimately progressed to the use of methamphetamine. The use of amphetamines exacerbates Anderson's mental deficits and impairments. Anderson has tried to overcome his addiction to methamphetamine, but without the support of his wife those efforts ultimately failed. Despite these serious impairments, Anderson had no history of criminal violence prior to the murders in question. Likewise, his family considered him a loving man, who always cared for his family and children and worked hard to support them.

The court, in this case, concluded that such evidence "serves to humanize a defendant and explain why an otherwise kind and loving family man can come to participate in a violent, murderous event."

Whether neuroscientific evidence will, or should be, similarly and increasingly accepted as valid, reliable, and relevant in like cases—as well as at other stages of the criminal justice process—continues to be evaluated and debated.

Key Terms

Actus Reus: A "bad act." Generally, one prong of what is required to be arrested, charged, and convicted of a crime. Associated with intention, or willful omission when a duty to act exists.

Aggravating Factors: Evidence that suggests a harsh sentence, as allowed by law, is warranted. Typically based upon criminal history, heinousness of the crime, victim type, or risk of future dangerousness.

"Double-Edged Sword": The proposition that evidence of brain abnormalities submitted to mitigate culpability for engagement in criminal behavior will also suggest that the individual has a neurological propensity for future dangerousness.

Dusky Standard: The two-prong standard set by the Supreme Court for a defendant's competency to stand trial that requires a defendant to have sufficient present ability to consult with his lawyer with a reasonable degree of rational understanding and a rational as well as factual understanding of the proceedings against him.

182　　　　　　　　　　　　　　　　　　　Neurocriminology

In Camera: A private hearing before the judge outside the presence of the jury and others.

Mens Rea: Literally "a guilty mind." Generally, one prong of what is required to be arrested, charged, and convicted of a crime.

Mitigating Factors: Any evidence that might suggest diminished culpability and result in reduced charges or a lesser sentence.

Strickland Claims: Stemming from the courts ruling in *Strickland v. Washington*, Strickland claims are a defendant's right to raise a claim of ineffective counsel. In relation to neuroscientific evidence, most often raised for counsel's failure to acquire, investigate, or present neuroscientific evidence that might mitigate a defendant's culpability and result in a lessor sentence.

Use Your Brain

Test Your Knowledge

1. Competency is typically predicated upon a defendant's:
   a. Lack of severe mental illness
   b. Present ability to consult with his or her lawyer with a reasonable degree of rational understanding
   c. A rational as well as factual understanding of the proceedings against him or her
   d. All of the above
   e. b and c

2. In general, neuroscientific evidence is admissible during the guilt–innocence phase of the trial if it is:
   a. Valid
   b. Relevant
   c. Reliable
   d. All of the above

3. To date, neuroscientific evidence is most often introduced in the guilt–innocence phase of criminal proceedings.
   a. True
   b. False

4. When entered during the sentencing phase of the criminal justice process, neuroscientific evidence is most often introduced as:
   a. A mitigating factor
   b. An aggravating factor
   c. Equally as mitigating and aggravating factors
   d. Neither as mitigating or aggravating factors

Prosecution and Sentencing

5. Which statement is most accurate concerning the relevance of neuroscientific evidence in appeals of death penalty sentences on the basis of ineffective assistance of counsel?
   a. Such claims generally succeed if neuroscientific evidence is not presented.
   b. Such claims generally fail, but are more apt to succeed if counsel does not articulate a compelling strategy for precluding neuroscientific evidence when relevant.
   c. Such claims generally fail regardless of whether neuroscientific evidence is introduced or relevant.
   d. Neuroscientific evidence is not admissible at the penalty phase of criminal proceedings.

Apply Your Knowledge

1. For decades, qualified mental health professionals have been permitted to offer testimony regarding a defendant's state of mind at various stages of criminal proceedings. In what ways and at what stages of the criminal justice process might neuroscientific evidence provide additional value? In what ways is such evidence duplicative or potentially distracting?
2. To date, neuroscientific evidence cannot demonstrate a causal relationship between a brain abnormality and engagement in criminal behavior. Given this, analyze possible reasons why such evidence should be precluded from admission at trial. What reasons argue for its inclusion?

Answer Key:
1. (e) 2. (d) 3. (b) 4. (a) 5. (b)

Bibliography

Aharoni, E., Vincent, G., Harenski, C., Calhoun, V. Sinnott-Armstrong, W., Gazzaniga, M., & Kiehl, K., (2013). Neuroprediction of future rearrest. *Proceedings of the National Academy of Sciences, 110*(15). Retrieved from www.pnas.org/cgi/doi/10.1073/pnas.1219302110.

Anderson v. Marty Sirmons, No. 04-6397 (Feb. 21, 2007).

Denno, D. (2015). The myth of the double-edge sword: an empirical study of neuroscience evidence in criminal cases. *Boston College Law Review, 56*, 493.

Espinoza, F., Vergara, V., Reyes, D., Anderson, N., Harenski, C., Decety, J., Rachakonda, S., Damaraju, E., Rashid, B., Miller, R., Koenigs, M., Kosson, D., Harenski, K., Kiehl, K., & Calhoun, V. (2018). Aberrant functional network connectivity in psychopathy from a large (*N*=985) forensic sample. *Human Brain Mapping, 39*, 1–11.

184 Neurocriminology

Drope v. Missouri, 420 U.S. 162 (1975).
Dusky v. United States, 362 U.S. 402 (1960).
Furman v. Georgia, 408 U.S. 238 (1972).
Godinez v. Moran, 509 U.S. 389 (1993).
Gregg v. Georgia, No. 74-6257, 428 U.S. 153 (July 2, 1976).
Hartocollis, A. (2006). Nearly 8 years later, guilty plea in subway killing. *New York Times*. Oct. 11.
Lockett v. Ohio, 438 U.S. 586 (1978).
Morissette v. United States, 342 U.S. 246 (1952).
Nelson v. State, No. F05-00846 (Fla. 11th Cir. Ct. Dec. 3, 2010) (hearing on Oct. 22, 2010).
Ovalle, D. (2010). A grotesque crime, a novel explanation. *Miami Herald*. Dec. 12.
Ovalle, D. (2010). Novel defense helps spare perpetrator of grisly murder. *Miami Herald*. Oct. 31.
People v. Goldstein, 6 N.Y.3d 119 (2005), 843 N.E.2d 727, 810 N.Y.S.2d 100.
Penry v. Lynaugh, 832 F.2d 915 (5th Cir. 1987).
Schmerber v. California, 384 U.S. 757 (1966).
State of Florida v. Grady Nelson, No. F05-00846 (11th. Fla. Cir. Ct., Dec. 4, 2010).
Strickland v. Washington, 466 U.S. 668 (1984).
United States v. Duhon, 104 F. Supp. 2d 663 (W.D. La. 2000).
United States v. Kasim, No. 2:07 CR 56 U.S. Dist. LEXIS 89137 (N.D. Ind. Nov. 3, 2008).
United States v. Kasim, No. 2:07 CR 56. U.S. Dist. (N.D. Ind. Jan. 21, 2010).
United States v. Montgomery, 18 U.S.C. § 1201(a)(1).
United States v. Sandoval-Mendoza, No. 04-10118 (Dec. 27, 2006).

Neuroscience and Law International Applications

8

Wide differences of opinion in matters of religious, political, and social belief must exist if conscience and intellect alike are not to be stunted, if there is to be room for healthy growth.

Theodore Roosevelt

No matter how powerful, countries cannot rule the whole world. The world is ruled by brains, by justice, by morals and by fairness.

Abdullah of Saudi Arabia

Learning Objectives

1. Describe ways in which neuroscience is applied to criminal justice efforts in different nations.
2. Explore the possible impact of local socio–cultural–political influences on the adoption of neuroscience internationally.
3. Compare differences and similarities in the translation of neuroscience to criminal justice practices throughout the world.

Introduction

Law and science are influenced by context. Society, culture, history, experience, politics, and perspective impact if and to what extent the two are entwined or siloed. Despite the globalization of many sectors, criminal justice policies and procedures are largely geographically bound. Prioritization of rehabilitation or retribution, concerns for free will or determinism, and the extent to which individual liberty or collective security is granted precedence are predominantly decided at the national level.

The manner in which different countries apply neuroscience to criminal justice policies is instructive, and offers additional data on both its potential and limitations—at least at the current time.

As in the United States, the introduction of neuroscience in many countries—when extant—is relatively recent; it is uncertain if initial applications foreshadow future practices. The following chapter, though clearly not exhaustive, explores some illustrative international translations of neuroscience to crime.

185

186 Neurocriminology

England and Wales

A 2007 *Press Telegram* article reported that Britain's Home Office proposed mandating the use of MRIs on convicted pedophiles undergoing questioning by probation and parole officers. MRIs would replace voluntary submission to "old-fashioned polygraphs," and assist law enforcement to better assess risk for re-offense.

Scientists were quoted as advocating caution, and calling for debates regarding the ethical use of emerging technologies. Others warned against England becoming a "Minority Report society" referencing the 2002 Steven Spielberg movie in which technology is utilized to anticipate crime, leading to arrest and convict of individuals prior to engagement in criminal acts.

Concerns for the inappropriate translation of neuroscience to criminal justice matters notwithstanding, England and Wales—which have common law, adversarial systems similar to the United States—have seen an increase in the use of neuroscience in criminal cases. A comprehensive study conducted by researchers Paul Catley and Lisa Claydon found that neuroscientific evidence was introduced in 1% (204) of criminal appellate court cases between 2005 and 2012, a figure that likely understates the prevalence of introduction at trial for which data are not compiled (see Box 8.1).

The majority of cases in which neuroscientific evidence was introduced concerned sentence reduction. Such evidence was also introduced to appeal convictions, resist extradition, or by the prosecution to appeal a sentence alleged to be too lenient.

In more than 70% in cases against conviction and more than 52% in cases against sentence, the appeals were denied. When appeals were upheld, however, neuroscientific evidence was largely or partially attributed to the success.

BOX 8.1   LIMITED PERSPECTIVES

Analyses of the use of neuroscientific evidence in criminal justice systems throughout the world are compromised by several factors.

First, many countries do not maintain central databases on criminal proceedings, resulting in a lack of information or reliance upon accounts that reach the public domain. The latter are often remarkable rather than representative.

When central databases do exist, the data itself are biased by the type of information included (e.g., the ruling rather than the grounds on which it was rendered) or the decisions reported (e.g., appeal decisions only, or decisions limited to certain providences or courts).

(Continued)

Neuroscience and Law

> **BOX 8.1 LIMITED PERSPECTIVES (*Continued*)**
>
> Consequently, examples and surveys of the use of neuroscientific evidence must necessarily be viewed with caution; no definitive conclusions can be drawn regarding the use of neuroscientific evidence in criminal justice settings in any particular nation from the relatively small samples currently available for study.
>
> These limitations notwithstanding, it is realistic to presume that neuroscientific evidence is introduced in more proceedings than captured by available databases or in the public domain. Additionally—at least in the case of appeal decisions for those countries that subscribe to a common law system—the case examples offer precedent that may inform future directions for neuroscience in the criminal courts.

Canada

Consistent with many jurisdictions, research on the use of neuroscientific evidence in the Canadian criminal justice system suggests that it is most often introduced during the sentencing phase. A large-scale study by researcher Jennifer Chandler, for example, found that of cases in which neuroscientific evidence was a factor, 65% pertained to sentencing decisions. The remainder included assessment of risk, and determination of mental capacity or fitness.

Most cases in which neuroscientific evidence was introduced were for serious offenses: The majority (21%) were sexual offenses, followed by homicide (17%) and assault (15%). The balance of cases in which neuroscientific evidence was introduced included robbery, theft, and breaking and entering, drug-related offenses, vehicular-related offenses, and threats and intimidation.

Interestingly, the neuroscientific evidence in the Canadian criminal cases reviewed did not include the introduction of brain imaging scans per se but rather predominantly focused upon known brain–behavior correlates revealed through studies related to fetal alcohol syndrome, traumatic brain injury, and other specific dysfunctions identified through neuropsychological testing.

The Netherlands

The use of neuroscientific evidence in criminal proceedings in the Netherlands has been found at all stages of criminal proceedings. A comprehensive study conducted by Katy de Kogel reviewed 231 cases heard between 2000 and 2012 in which neuroscientific or genetic-behavioral information was introduced. In 90%, neuroscientific evidence was referenced independently, and in 94%, it was introduced in conjunction with genetic-behavioral evidence. In nearly one-third of cases (31%), the evidence was used to answer questions pertaining to diminished accountability. The second largest application of neuroscientific evidence was to determine intent (9%), followed by risk of recidivism and guilt versus negligence (6% each). The balance of cases utilized such evidence for a wide variety of applications, ranging from evidence of committing the offense, premeditation, duress, and excessive self-defense.

BOX 8.2   THE NEUROSCIENCE OF PREMEDITATION

In contrast to many jurisdictions, Dutch courts appear to be utilizing neuroscience more to inform a broader variety of questions that arise in criminal justice proceedings, including those related to premeditation, as exemplified by the case below:

District Court Amsterdam, March 28, 2008

In 2007, a 63-year-old man stabbed a friend nine times as a result of which she deceased. The suspect declared that he was annoyed by the victim's behavior. He furthermore declared that he saw that the victim lost a lot of blood as a result of the stabbing and furthermore lost her consciousness several times. When she regained consciousness and tried to get up, he stabbed her again. At the time of the incident, the suspect was intoxicated with alcohol and cocaine. According to the expert witness, a behavioral neurologist, the suspect's behavior during the incident was affected by damage to his frontal lobes. More specifically, the brain damage had rendered the suspect unable to control his impulses and reflect on his actions in difficult situations. The alcohol and cocaine were believed to have aggravated his impulsive behavior. Additionally, the expert witness stated that the suspect's brain damage interfered with his free will. The presiding judge decided that the suspect had acted intentionally. More specifically, he stated that although the suspect's behavior was affected by the frontal lobe damage, he did not lack complete insight into the consequences of his actions. According to the judge, the suspect was aware of the possibility that the victim would die as a result of the harm that he was inflicting on her. Consequently, the court decided that the suspect had committed the act of intentionally killing someone (manslaughter). On the basis of the expert witness' report, however, the judge decided that the suspect had severely diminished responsibility for his actions as a result of the frontal lobe damage, which eventually resulted in reduced sentencing of 18 months' imprisonment, plus detention during Her Majesty's pleasure.

Neuroscience and Law

189

India

> This case presents a tragic scenario as the budding and flourishing love relationship between accused No. 1 Aditi and deceased Udit Bharati, which was on the threshold of marriage and sailing smoothly with consent and approval of the parents on both sides, got swerved to the wrong side and sank into tragedy when accused No. 1—Aditi came in to contact with accused No. 2 Pravin, got enamored by him, fell in love with him and left the deceased to settle with Pravin and ultimately, as per prosecution case, eliminated deceased from this world, in conspiracy with Pravin, for which both of them are indicted and charge sheeted...

So begins the 2007 Judgment of the State of Maharashtra, India against 24-year-olds Aditi Baldev Sharma and Pravin Premswarup Khandelwal. The State alleged that, after secretly marrying, the two went on to formally end Sharma's family-arranged engagement to Bharati through her covert poisoning of him to death with arsenic-laced food. The State's investigation revealed circumstantial evidence that included traces of arsenic in Sharma's purse, phone records that confirmed a call she made to Bharati to schedule the meeting at which she allegedly feed him arsenic-laced food, and hotel records that confirmed she and Khandelwal stayed at a hotel nearby the location where Bharati was allegedly murdered.

In her defense, Sharma suggested that Bharati was distraught over her love for Khandelwal and committed suicide. Khandelwal maintained that he was uninvolved in the murder.

The two voluntarily underwent polygraph examinations. Sharma also agreed to undergo a Brain Electrical Oscillations Signature—BEOS—test. BEOS was developed by Indian neuroscientist Champadi Raman Mukundan based upon Brain Fingerprinting EEG technologies advanced by American neuroscientists including Lawrence Farwell.

For the test, 32 electrodes were placed on Sharma's head. She remained silent while, during the course of an hour, investigators read statements regarding the crime ("I bought arsenic") as well as neutral statements ("The sky is blue"). The brain scans were said to be able to identify conceptual versus experiential knowledge; the latter would demonstrate memory for the event.

At the time of Sharma and Khandelwal's trial, the BEOS test had been utilized as part of the Indian criminal justice process an estimated 75 times.

Theirs was the first, however, in which its use served as a basis for conviction; Sharma was found guilty of murder and Khandelwal was found guilty of conspiracy to commit murder.

In his opinion, the presiding judge defended the BEOS test, citing it in as part of the evidence that implicated Sharma: "In the Polygraph and BEOS

190   Neurocriminology

Test, responses of Aditi to the set of questions put to her were found to be deceptive and proving her experiential knowledge, respectively."

Both Sharma and Khandelwal maintained their innocence.

The international neuroscientific community, in general, responded with disfavor to India's decision to translate the technology to determine guilt. A *New York Times* article that appeared in the months after the verdict quoted several neuroscientists, including J. Peter Rosenfeld—a psychologist and one of the developers of initial electroencephalogram-based lie detection systems—as stating,

> Technologies which are neither seriously peer-reviewed nor independently replicated are not, in my opinion, credible. The fact that an advanced and sophisticated democratic society such as India would actually convict persons based on an unproven technology is even more incredible.

Within 3 months following the convictions, India's National Institute of Mental Health and Neuro Sciences (NIMHNS) echoed this perspective, issuing a declaration recommending the discontinuation of BEOS for evidentiary purposes, citing the lack of scientific peer review.

Sharma subsequently appealed her conviction and was released on bail 6 months later.

**BOX 8.3   BRAIN ELECTRICAL OSCILLATIONS
SIGNATURE: MORE THAN A DECADE LATER**

More than 10 years after its instrumental—and controversial—role in the murder and conspiracy to commit murder convictions of Aditi Baldev Sharma and Pravin Premswarup Khandelwal, the BEOS technology developed by Dr. Mukundan has continued to evolve and continues to be used in India's criminal justice process. Offered by Axxonet, the test now utilizes a "new set of algorithms" known as the Neuro Signature System (NSS). According to the Axxonet website (www.axxonet.com/pdfs/NSS.pdf), the test has been used in over 700 criminal cases since it was introduced in 1999.

Italy

Two cases within the public domain demonstrate the Italian criminal justice system's use of neuroscientific evidence.

The first case involved Abdelmalek Bayout who, in 2007, confessed to the murder of a man who insulted him for wearing eye makeup, which Bayout stated he wore for religious reasons. During trial, the defense produced expert witnesses who testified that Bayout was mentally ill and was

Neuroscience and Law

not taking his psychotropic medications at the time of the murder. The judge agreed Bayout's mental status was a mitigating factor in his crime and sentenced him to 9 years and 2 months in prison instead of the 12 years he would have otherwise received.

In 2009, Bayout was granted an appeal hearing. Cognitive neuroscientific tests, including an MRI, were conducted. Based upon abnormalities found in the neuroimaging scan, as well as in levels of his neurotransmitter metabolizing enzyme monoamine oxidase A—the so-called "Warrior Gene" that had been previously linked with aggression and criminal behavior in some studies—the appeal judge reduced Bayout's sentence by an additional year.

The case represented the first time that a European court utilized a genetic-behavioral correlation as the basis for a sentence reduction.

Two years later, the Italian courts did so for a second time. This time, MRI and low MAOA gene activity evidence was the basis for reducing the sentence of a convicted murderer from life to 20 years.

The case involved Stefania Albertani, who was convicted of murdering her sister by force feeding her psychotropic medication before setting her corpse on fire.

Documents filed on appeal included the MRI scan of Albertani's brain as well as those of 10 healthy female control subjects. In comparison to the healthy controls, Albertani showed gray matter reductions in the anterior cingulate gyrus and the insula, associated with behavioral regulation and aggression, respectively. Genetic testing also showed that Albertani had low MAOA gene activity.

In reducing her sentence, the judge found that the neuroscientific evidence showed that Albertani was "partially mentally ill."

BOX 8.4   SHARED SCIENTIFIC SENSIBILITIES

Although laws are jurisdictional, science and its ethical practice is greatly influenced by universal guidelines and standards. On the international level, these guidelines have been propagated by the International Bioethics Committee (IBC) of the United Nations Educational, Scientific and Cultural Organization (UNESC). Founded in 1993, the IBC is comprised of 36 independent experts from Belgium, Brazil, Canada, Cuba, Czech Republic, Democratic Republic of Congo, Egypt, Finland, France, Gabon, Germany, Ghana, Islamic Republic of Iran, Italy, Jamaica, Kenya/South Africa, Kuwait, Latvia, Lebanon, Malaysia, Mexico, the Netherlands, Norway, Panama, Poland, Republic of Korea, Romania, Senegal, Singapore, Spain, Sri Lanka, Switzerland, Thailand, Tunisia, United States of America, and Uruguay.

(Continued)

192 — Neurocriminology

> **BOX 8.4 SHARED SCIENTIFIC SENSIBILITIES (*Continued*)**
>
> In 2005, the IBC published a Universal Declaration on Bioethics and Human Rights. The nonbinding declaration does not explicate the translation of neuroscience to law. It does, however, adopt the following as a general principle: "In applying and advancing scientific knowledge, medical practice and associated technologies, direct and indirect benefits to patients, research participants and other affected individuals should be maximized and any possible harm to such individuals should be minimized." The IBC further called for "opportunities for informed pluralistic public debate, seeking the expression of all relevant opinions" in relation to decision-making concerning translational science.
>
> In 2006, an interdisciplinary group of neuroscientists, scholars, clinicians, lawyers, and ethicists founded the International Neuroethics Society (INS) "to promote the development and responsible application of neuroscience through interdisciplinary and international research, education, outreach and public engagement for the benefit of people of all nations, ethnicities, and cultures." As part of its efforts, INS responded to the call for comments published in the U.S. Federal Register (January 31, 2014) for the Presidential Commission for the Study of Bioethical Issues. The INS listed 12 areas that it deemed important for consideration, choosing to provide additional detail on five "due to their rapid advancement and the immediate need for more government and public consideration of the ethical impact on society." Among the five was an area that INS titled "Responsibility, Moral Agency and the Law," which included "the use of neuroscientific arguments and neuroimaging in a legal context."
>
> The cases summarized and highlighted in this chapter represent those for which data are available; undoubtedly, they represent but a sample of the instances in which neuroscience is applied to criminal justice proceedings worldwide, suggesting international consideration of these issues is indeed an appropriate domain for informed debate and prioritization.

Use Your Brain

Test Your Knowledge

1. As in the United States, studies on the use of neuroscientific evidence in Canada, England, and Wales have found that it is most often introduced in which aspect of the criminal justice process:
   a. Competency
   b. Guilt

Neuroscience and Law

c. Sentencing
d. Mental capacity

2. Research on appeals decisions in England and Wales found that, in the majority of cases, the introduction of neuroscientific evidence resulted in amendments to prior decisions related to convictions or sentences.
   a. True
   b. False

3. In a controversial ruling that was subsequently appealed, an Indian court used the Brain Electrical Oscillations Signature (BEOS) test to determine that a defendant:
   a. Was insane at the time of the offense
   b. Did not commit the crime
   c. Had experiential memory of the crime
   d. Was psychopathic

4. In two publicized cases, Italian courts reduced the sentences of convicted murderers based upon MRI and genetic testing.
   a. True
   b. False

5. Assessments on the use of neuroscientific evidence internationally must be viewed with caution due in part to:
   a. Database biases
   b. Misrepresentation by the media
   c. Lack of a consistent definition for "neuroscientific"
   d. None of the above

Apply Your Knowledge

1. In England and Wales, neuroscientific evidence did not impact appeal decisions in the majority of cases in which it was introduced. In the two Italian cases presented in this chapter, neuroscientific evidence did influence appeal decisions. On what bases, if any, should neuroscientific evidence affect the outcome of criminal appeals? State your rationale.

2. In India, the use of the BEOS test to determine guilt was met with international concern, and resulted in a declaration against its use by India's National Institute of Mental Health and Neuro Sciences. Should there be a "global" standard for the use of neuroscience in criminal justice proceedings? What might be some advantages and disadvantages of an international standard?

Answer Key:
1. (c) 2. (b) 3. (c) 4. (a) 5. (a)

Bibliography

Abdullah of Saudi Arabia Quotes (n.d.). BrainyQuote.com. (2018, February 4). Retrieved from BrainyQuote.com website: www.brainyquote.com/quotes/abdullah_of_saudi_arabia_542922.

Catley, P., & Claydon, L. (2015). The use of neuroscientific evidence in the courtroom by those accused of criminal offenses in England and Wales. *Journal of Law and the Biosciences, 2*(3), 510–549. doi:10.1093/jlb/lsv025.

Chandler, J. (2015). The use of neuroscientific evidence in Canadian criminal proceedings. *Journal of Law and the Biosciences, 2*(3), 550–579. doi:10.1093/jlb/lsv026.

Feresin, E. (2009). Lighter sentence for murderer with "bad genes": Italian court reduces jail term after tests identify genes linked to violent behaviour. *Nature.* doi:10.1038/news.2009.1050.

Giridharada, A. (2008). India's Novel Use of Brain Scans in Courts is Debated. *New York Times.* Sept. 14.

Hunter, G. (2013). *Selected speeches and writings of Theodore Roosevelt.* New York: Vintage Books.

Klaming, L., & Koops, E. J. (2012). Neuroscientific evidence and criminal responsibility in the Netherlands. In T.M. Spranger (Ed.), *International Neurolaw: A comparative analysis* (pp. 227–256). Heidelberg: Springer.

Owens, B. (2011). Italian Court Reduces Murder Sentence Based On Neuroimaging Data. *Nature.com.* Sept. 01.

State of Maharashtra, through Nigdi Police Station vs. 1] Aditi Baldev Sharma 2] Pravin Premswarup Khandelwal, Sessions Case No. 508/07.

Neurocriminology
Present Context and Possible Future

9

Nature is full of infinite causes
Leonardo da Vinci

In the decades since *The Economist* warned that "brain science" could "gut the concept of human nature," neuromodesty has generally prevailed.

Neuroscience has neither healed nor undermined the criminal justice system.

As was true of past approaches to addressing crime, the socio–cultural–political context in which neurocriminology is developing has and will likely continue to influence its application.

Interdisciplinary collaboration, facilitated by technology, higher education, and investments such as those made by the MacArthur Foundation Research Network on Law and Neuroscience and the Presidential Commission for the Study of Bioethical Issues, has facilitated both significant research on the application of neuroscience to crime, and robust dialogue exploring its many ramifications.

The globalization of communication has at once broadened opportunities for innovation and served to protect against "neuro-overreach." When neuroscience has been applied prematurely or badly, as when EEG data were admitted as evidence of murder, or simply misunderstood, as when colorful MRI images are introduced to prove the cause of a defendant's criminal behavior, the scientific and legal communities respond, and corrective action generally follows.

The feared "CSI effect" of neuroscientific technology inappropriately biasing judicial outcomes has not taken hold. In addition to the success of educational efforts by prominent researchers in the field, a proliferation of "true-crime" dramas as well as the ubiquitous application of neuroscience in other contexts, such as the medical field, has likely tempered the "whiz-bang" impact on an informed public.

Previous scientific advances precipitated legal decisions that have also supported prudence in relation to neurocriminology. Established standards for the admissibility of scientific evidence that require consideration of validity, reliability, and relevance have served to ensure that when neuroscience is translated to the criminal justice system, its current limitations—such as small and nonrepresentative sample groups, the impossibility of determining brain state at time of offense, the inability to make causal links, and the inability to definitively apply general research findings to specific individuals—are recognized.

Despite its limitations, neuroscience has played an edifying role in several criminal justice contexts, such as in revisiting the constitutionality of sentencing adolescents to death or to life in prison, and in explaining mitigating factors such as the impact of traumatic brain injury on culpability.

Neuroscientific advances that more closely mimic "mind reading"—such as in the areas of lie detection and memory retrieval—have not yet proven reliable. To date, replicating significant study results in these domains has proven elusive. And yet intelligent discussions are already underway regarding the need to safeguard "cognitive liberty" if—or more likely when—these technologies are available.

Nor has neuroscience undermined or resolved the free will-determinism divide. To the contrary, the current state of neuroscientific research offers evidence that could be appropriated by each philosophical position: Although not causal, neuroscience has provided strong evidence of brain–behavior correlates, suggesting "will" is no more than a series of neurochemical processes. It has also demonstrated that there are behavior–brain correlates, as certain volitional behaviors—both adaptive and maladaptive—can alter brain functioning; under certain conditions, we seem able to will ourselves to be neurochemically different.

The brain's current refusal to yield unambiguous responses has precluded neuroscience from—as hoped or feared—providing the categorical answers sought by the criminal justice system: Competency/Incompetent. Truthful/Lying. Sane at the time of the Offense/Insane at the time of the Offense. Innocent/Guilty. As a 2015, the Gray Matters report issued by the Presidential Commission for the Study of Bioethical Issues stated, "Neuroscience cannot answer central normative questions that are important to society—for example, why we punish criminals and what it means to be a morally responsible or free human being."

Instead, neuroscience addresses different but related complexities. Over the last several decades, the field has engaged in an unprecedented exploration of the structural and functional abnormalities associated with various behaviors, including criminal activity. The findings amassed during this relatively brief period suggest vital advances will continue in the near future and beyond.

Recent events have offered a sobering reminder of the continued, urgent need for efforts that can successfully prevent, and contain criminal violence.

In February 2018, results from the brain examination of the man responsible for the worse mass shooting in U.S. history were released. News sources quoted Dr. Hannes Vogel, director of neuropathology at Stanford University

Neurocriminology: Present Contex and Possible Future

as stating, "With a good deal of screening, I didn't see anything" that could explain why Las Vegas gunman Stephen Paddock committed his crime.

Dr. Vogel reported finding corpora amylacea (CA) in the hippocampus and the frontal lobes of Paddock's brain. He indicated that "Most people would have them at that age, but not in that profusion."

The origin and potential function of CA remains largely unknown. Consistent with Dr. Vogel's reports, researchers have concluded that low numbers of CA are commonly detected in the aging brain of normal individuals. Higher levels of CA in the brain have been associated with neurodegenerative diseases, particularly Alzheimer's disease, multiple sclerosis, and epilepsy—diseases that have not been associated with the violence that Paddock exhibited.

Dr. Vogel expressed hope that the public would be reassured that Paddock's doctors had not missed diagnosing a tumor or other brain disorder that could have been treated.

Victims and witnesses to the crime were left to wonder why an individual who seemingly offered so few obvious predictors of violence went on to engage in one of the nation's deadliest examples of it.

In the same month that the Paddock brain evaluation report was released, the deadliest U.S. school shooting since the 2012 massacre at Connecticut's Sandy Hook Elementary School occurred. On Valentine's Day, 19-year-old Nikolas Cruz shot and killed 17 with an AR-15 rifle at the Marjory Stoneman Douglas High School, from which he had been expelled the year before. Cruz was apprehended following the shooting. Reports of his extensive history of behavioral problems, which included a history of bullying, vandalism, hurting animals, and posting threatening messages online, were rampart.

The lack of effective intervention to prevent a reportedly very disturbed child and teenager from becoming a mass murderer is incomprehensible.

Psychologist and neuroscientist Joshua Buckholtz and legal scholar David Faigman have eloquently and accurately written that neuroscience's impact on the legal system "often sheds more heat than light…"

The promise of neuroscience is that, in time, it will forge new paths to predict and interrupt or—at the least—more precisely explain, criminal behavior. Data will clearly not preempt society's right or responsibility to decide what to do with those who demonstrate propensity for or engagement in criminal behavior. But advances in neurocriminology can provide significantly more—and more effective—options for dealing with those who do so.

Bibliography

Buckholtz, J., Faigman, D. (2014). Promises, promises for neuroscience and law. *Current Biology*, Sep 22; *24*(18):R861–R867. doi:10.1016/j.cub.2014.07.057.

Fink, S. (2018). Las Vegas Gunman's Brain Exam Only Deepens Mystery of His Actions. *New York Times*. Feb. 9.

Presidential Commission for the Study of Bioethical Issues (2015). *GRAY MATTERS: Topics at the Intersection of Neuroscience, Ethics, and Society*, Vol. 2. Mar. 2015. Retrieved from: www.bioethics.gov.

Index

A

Abnormal object perception, 41
Absence seizures, 49
Acalculia, 41
Acetylcholine (ACh), 35
ACh. *see* Acetylcholine
"Acquisitiveness," 5
Actus reus, 164, 181
Adolescent murderers, 98
Affective murderers, 95–97
Affective violence, 95, 106
Aggravating factors, 172, 181
Agnew, Robert, 12
Agnosia, 41
Agonists, 36, 56
Agraphia, 41
Akers, Ronald, 10
Alzheimer's disease, 50
Amino acid tyrosine, 36
Amygdala, 43, 57
Anomie theorists, 11
Anoxia, 47
ANS. *see* Autonomic nervous systems
Antagonist, 36, 56
Arterial spin labeling (ASL), 71
ASL. *see* Arterial spin labeling
Astrocytomas, 45
Atonic seizures, 49
Aura stage, 50
Autonomic nervous systems (ANS), 21, 34
Axial plane, 65, 75
Axons, 34

B

BAC levels. *see* Blood alcohol content levels
Balint's syndrome, 41
Basal ganglia, 38, 56
Bayes, Thomas, 2
Beccaria, Cesare, 5
Bennett, William, 19
Bentham, Jeremy, 3
    panopticon, 4

BEOS test. *see* Brain Electrical Oscillations
    Signature test
Blood alcohol content (BAC) levels, 53
Blood oxygenation sensitive steady-state
    (BOSS), 71
Blood oxygen level-dependent (BOLD)
    measurements, 71
BOLD measurements. *see* Blood
    oxygen level-dependent
    measurements
BOSS. *see* Blood oxygenation sensitive
    steady-state
"Bottom–up processing," 42
Brain
    cord tumors, 46
    and drug abuse, 125–126
    dysfunction
        epilepsy, 49–50
        neurocognitive disorder, 50–51
        personality disorders, 52–53
        psychotic and mood disorders, 51–52
        stroke, 46
        substance abuse, 53–55
        trauma, 47–49
        tumor, 45–46
    and early-onset antisocial behaviors,
        122–123
    and justified *versus* unjustified homicide,
        96–97
    plasticity, 121
    structure and functions of
        Brodmann's areas, 44
        cerebral cortex, 38–43
    tumors, 45
Brain Electrical Oscillations Signature
    (BEOS) test, 189
Broca's area, 41
Brodmann's areas, 44
Brunner, Hans, 21

C

CA. *see* Corpora amylacea
California Penal Code, 132–136

Index

Canada, 187
Carter, Jimmy, 17
CAT. *see* Computerized axial tomography
Catecholamines, 36
CBCA. *see* Criteria-Based Content Analysis
CBF. *see* Cerebral brain flow
Central nervous system (CNS), 34, 55
Cerebral brain flow (CBF), 67–68
Cerebral cortex
   amygdala, 43
   cingulate cortex, 43
   frontal lobes, 41–42
   hippocampus, 42–43
   hypothalamus, 42
   limbic system, 42
   occipital lobes, 39–40
   parietal lobes, 40–41
   temporal lobes, 40
Cerebrospinal fluid, 45
Cerebrum, 38–39, 56
Chicago area project, 11
Chronic traumatic encephalopathy (CTE), 47
Cingulate cortex, 43, 57
Classical School of Criminology, 2–5, 23
Clonic seizures, 50
Clustered Personality disorders, 52
CNS. *see* Central nervous system
Cocaine, 55
"Combativeness," 5
Computed tomography (CT), 66
Computerized axial tomography (CAT), 66
Conservative criminology, 9–19, 24
Continuous performance test (CPT), 94, 106
Control theories of crime, 13–19, 24
Contusion, 47
Coronal plane, 65, 75
Corpora amylacea (CA), 197
Corpus callosum, 38, 43, 56
Counsel, ineffective use of, 180
CPT. *see* Continuous performance test
Crampton, Peter, 23
Cranium, 45
Criminal investigation, 137–152
Criminal justice system
   prevention and investigation, 111–152
   prosecution and sentencing, 159–181
Criminals, in lab
   murderers
      adolescent, 98

      non-homicidal offenders, 98–101
      predatory *versus* affective murderers, 95–97
      with severe mental illness, 97–98
   sexual offenders
      pedophiles, 104–105
      rapists, 101–104
      subtype, 105–106
      white-collar criminals, 105
Criteria-Based Content Analysis (CBCA), 138, 139
Critical criminology, 9–19, 24
CT. *see* Computed tomography
CTE. *see* Chronic traumatic encephalopathy

D

Daubert standard, 148, 152
Declarative memory, 33, 152
Delusional disorder, 51
Dendrities, 34
Diencephalon, 38
Diffusion tensor imaging (DTI), 67
Diminished culpability, 117, 152
DLPFC. *see* Dorsolateral prefrontal cortex
Dopamine, 36
Dopaminergic neurons, 36
Dopaminergic system, 127, 153
Dorsolateral prefrontal cortex (DLPFC), 41, 123
Double-edged sword, 175–176, 181
Downward departure, 118, 152
Drug
   abuse, 125–126
   brain, 127–130
   courts, 125
DSC. *see* Dynamic susceptibility contrast
DTI. *see* Diffusion tensor imaging
Dusky standard, 160, 181
Dynamic susceptibility contrast (DSC), 71
Dyscalculia, 41
Dysgraphia, 41

E

Early-onset antisocial behaviors, 122–123
EEG. *see* Electroencephalography
Electroencephalography (EEG), 68–70
Electrophysiological techniques, 68
England, 186–187

Index

201

Ependymal cells, 36
Epilepsy, 49–50
Evoked potentials, 68–70
External granular layer of cerebral
    cortex, 39
External pyramidal layer of cerebral cortex,
    39

F

Ferri, Enrico, 8
Finger agnosia, 41
fMRI. *see* Functional magnetic resonance
    imaging
Focal seizures, 50
Forebrain, 38, 56
Forensic neuropsychological/psychological
    assessments, 64–65
Frontal lobes, 41–42, 57
Frontotemporal dementia (FTD), 50–51,
    100–101
Frye standard, 148, 153
FTD. *see* Frontotemporal dementia
Functional magnetic resonance imaging
    (fMRI), 70–71
Functional neuroimaging techniques
    electroencephalography/
        evoked potentials and
        magnetoencephalography/
        magnetic source imaging,
        68–70
    functional magnetic resonance imaging,
        70–71
    positron emission tomography, 71
    single-photon emission computerized
        tomography, 71–72
Fusiform layer of cerebral cortex, 39

G

G2i counterargument, 87
GABA. *see* Gamma-aminobutyric acid
Gall, Franz Joseph, 5
Gamma-aminobutyric acid (GABA), 35
Garofalo, Raffaele, 8–9
Gertsmann's syndrome, 41
Gigante, Vincent, 87–90
Glial cells, 36, 55
Gottfredson, Michael, 14
Graham, Terrance, 117–119
Gray matter, 34, 38–39, 56
Guilt phase of criminal process, 164–171

H

Hemorrhagic strokes, 46
Hinckley, Jr., John, 83–87
Hindbrain, 38
Hippocampus, 42–43, 57
Hirschi, Travis, 14
Homicidal adolescents, 98
Hypoactive autonomic nervous systems, 21
Hypothalamus, 38, 42, 56, 58

I

Ictus stage, 50
In camera, 169, 170, 182
India, 189–190
Infant neurons, 121
Inferential gaps, 73–74
Infragranualar layers of cerebral cortex, 39
Insanity Defense Reform Act of 1984, 86, 90
Internal granular layer of cerebral cortex, 39
Internal pyramidal layer of cerebral
    cortex, 39
Interneurons, 35
Intracranial hematoma, 47
Italy, 190–192

J

Jackson, Kuntrell, 119
Johnson, Eric, 152
Justified homicide, 96–97

K

Kentucky v. Stanford, 113–114
Klüver-Bucy syndrome, 43

L

Language disorder, 41
Lauterbur, Paul, 66
Left realism, 9–19, 24
Legislating leniency, 132–136
Lie detection, 146–152
Limbic system, 42, 57
Lombroso, Cesare, 7–8

M

Magnetic resonance imaging (MRI), 66–67
Magnetic source imaging, 68–70

202   Index

Magnetoencephalography (MEG), 68–70
Mainstream criminology, 10–13
Mansfield, Peter, 66
MAOA. *see* Monoamine oxidase A
Matza, David, 13
MEG. *see* Magnetoencephalography
Meloy, Reid, 95
Meningiomas, 46
Mens rea, 164, 182
Merton, Robert, 11–12
Mesencephalon, 38
Methamphetamine, 55
Microglia, 36
Midbrain, 38, 56
"Military policing" police, 17
Miller, Evan, 119–120
Mitigating factors, 172, 174–175, 182
Molecular layer of cerebral cortex, 39
Monoamine oxidase A (MAOA), 21–22
Mood disorders, 51–52
MRI. *see* Magnetic resonance imaging
Multiform layer of cerebral cortex, 39
Multivoxel pattern analysis (MVPA),
    144, 153
Murderers
    adolescent, 98
    non-homicidal offenders, 98–101
    predatory *versus* affective murderers,
        95–97
    with severe mental illness, 97–98
MVPA. *see* Multivoxel pattern analysis
Myelin, 36, 55
Myoclonic seizures, 50

N

Neocortex, 39
Netherlands, The, 187–188
Neurocognitive disorder, 50–51
Neuroimaging, 164–165
    bias, 89, 90
    functional techniques, 67–72
    strengths and limitation of, 72–75
    structural techniques, 66–67
    works, 65–66
Neurons, 34–35, 55
Neuroplasticity, 121–122, 153
Neuroprediction of dangerousness,
    177–178
Neuroscience, 137–152
    and law, 185–192

Neurotransmitters, 34, 55
Non-homicide
    adolescents, 98
    offenders, 98–101
Norepinephrine, 36

O

Occipital lobes, 39–40, 56
OFC. *see* Orbitofrontal cortex
Oligodendrocytes, 36
Oligodendrogliomas, 46
Orbitofrontal cortex (OFC), 96–97

P

Parahippocampal gyrus, 97, 101–102, 106
Parietal lobes, 40–41, 57
Parkin, Chris, 23
Partial seizures, 50
PDs. *see* Personality disorders
Pedophilia, 102–105
Peripheral nervous system (PNS), 34, 55
Personality disorders (PDs), 52–53
PET. *see* Positron emission tomography
PFC. *see* Prefrontal cortex
Phrenology, 5–9
Planck, Max, 9
PNS. *see* Peripheral nervous system
Positivist school of criminology, 7, 23
Positron emission tomography (PET), 71
Post-traumatic stress disorder (PTSD),
    130–131
Pre-Classical Period, 2
Predatory murderers, 95–97
Predatory violence, 95, 106
Prefrontal cortex (PFC), 41–42
Prodormal stage, 50
Prosencephalon, 38
Psychological science, 138–140
Psychopathy, 52–53
Psychotic disorders, 51–52
Psychotropic medications, 36
PTSD. *see* Post-traumatic stress disorder

Q

qEEG. *see* Quantitative
    electroencephalography
Quantitative electroencephalography
    (qEEG), 69–70

Index

R

Rapists, 101–104
RAS. *see* Recticular activating system
rCBF. *see* Regional cerebral brain flow
Real-time-functional magnetic resonance imaging, 126, 153
Recidivism, 15
Reckless, Walter, 13
Regional cerebral brain flow (rCBF), 71
Reynolds, Morgan, 19
Rhombencephalon, 38
Right orbitofrontal tumor, 102–104
Rousseau, Jean-Jacques, 2
Rubenstein, Jacob, 82–83
Ruby, Jack, 82–83

S

Sagittal plane, 65, 75
Samenow, Stanton, 19
SAMHSA. *see* Substance Abuse and Mental Health Services Administration
Schizophrenia, 51–52
Seminal cases, 30–33
Sentence mitigation, 172–181
Serotonin, 36
Severe mental illness, murderers with, 97–98
Sexual offenders
    pedophiles, 104–105
    rapists, 101–104
Simmons, Roper v., 114–117
Single-photon emission computerized tomography (SPECT), 71–72
Skull fracture, 47
Social bond theory, 14
Sociopolitical context, 17
Somatosensory cortex, 40
Specialty courts, 136–137
SPECT. *see* Single-photon emission computerized tomography
Spinal cord tumors, 46
SQUIDS. *see* Superconducting Interference Devices
Strain theorists, 11
Strickland claims, 180, 182
Stroke, 46
Structural neuroimaging techniques
    computed tomography/computerized axial tomography, 66

magnetic resonance imaging, 66–67
Suboptimal arousal theory, 21
Substance abuse, 53–55
Substance Abuse and Mental Health Services Administration (SAMHSA), 53
Superconducting Interference Devices (SQUIDS), 68–69
Sutherland, Edwin, 10
Sykes, Gresham, 13
Sylvian fissure, 41

T

TBI. *see* Traumatic brain injury
Telencephalon, 38
Temporal lobes, 40, 56–57
Thalamus, 38, 56
Tomography, 75
Tonic–clonic seizures, 50
Tonic seizures, 49
"Top–down processing," 41
Trauma, 47–49
Traumatic brain injury (TBI), 47
Tumor, 45–46

U

Unjustified homicide, 96–97

V

Ventromedial prefrontal cortex (VMPFC), 41
Veterans Treatment Court (VTC), 130–131
VMPFC. *see* Ventromedial prefrontal cortex
VTC. *see* Veterans Treatment Court

W

Wales, 186–187
Warrior genes, 21–23
Wernicke's area, 40
White matter, 34, 56
Whitman, Charles Joseph, 80–81
Witness
    memory recovery, 144–146
    problem, 143
Writing disability, 41